Unified Vision for a Sustainable Future

Mir Sayed Shah Danish
Editor

Unified Vision for a Sustainable Future

A Multidisciplinary Approach Towards the Sustainable Development Goals

Springer

REPA
Research and Education Promotion Association

Editor
Mir Sayed Shah Danish
Energy Systems (Chubu Electric Power) Funded Research Division
Institute of Materials and Systems for Sustainability (IMaSS),
Nagoya University
Nagoya, Aichi, Japan

ISBN 978-3-031-53573-4 ISBN 978-3-031-53574-1 (eBook)
https://doi.org/10.1007/978-3-031-53574-1

This Springer imprint is published by the registered company Springer Nature Switzerland AG
The registered company address is: Gewerbestrasse 11, 6330 Cham, Switzerland

Paper in this product is recyclable.

Preface

Imagine a world where sustainability is not just a goal but a reality. This is the vision that drives the 2024 International Conference on Collaborative Endeavors for Global Sustainability (CEGS) at the University of British Columbia in Vancouver, Canada.

Under the theme "Unified Vision for a Sustainable Future: A Multidisciplinary Approach Towards the SDGs," we bring together minds from technology, policy, and practice to tackle the sustainability challenges of the twenty-first century. Our impressive 62% acceptance rate has culminated in a collection of research papers that are not just academic contributions but beacons of hope and innovation.

What sets CEGS-2024 apart is our commitment to turning theory into action. Our sessions extend beyond academic discussion to interactive workshops and collaborative projects, all aimed at moving us closer to the SDGs.

These papers, aligning with the conference agenda of driving impactful change, deal with the intersections of technology, policy, and practical implementation within the sustainability domain. It started with "Data-Driven Pathways to Sustainable Energy Solutions," exploring how data analytics can revolutionize energy efficiency and sustainability. Following this, "Multidimensional Analysis and Optimization of Bus Loads for Enhanced Renewable Energy Integration in Power Systems" showcases how renewable energy can be better integrated into power systems, emphasizing the importance of renewable resource integration in the form of ultimate solutions in our sustainable future.

The third paper, titled "An Overview of Inverter and Converter Roles in Microgrids," dissects the technical aspects crucial for the functioning of modern sustainable energy systems. The fourth paper in this proceeding book, "Integrating Machine Learning in Energy Systems: A Techno-Economic Framework for Enhancing Grid Efficiency and Reliability," presents a cutting-edge perspective on enhancing grid efficiency and reliability, highlighting the indispensable role of artificial intelligence in future energy solutions. "Renewable Energy and Power Flow in Microgrids: An Introductory Perspective" then provides an introductory outlook on the potential of microgrids in revolutionizing energy distribution.

The sixth paper in this series, "Sustainable Energy Policy Formulation through the Synergy of Backcasting and AI Approaches," proposes innovative strategies for policy formulation, blending futuristic visioning and AI. Finally, "A Blueprint for Sustainable Electrification through Designing and Implementing PV Systems in Small Scales" presents practical insights into the implementation of photovoltaic systems, underscoring the importance of small-scale solutions in the sustainability puzzle.

As we gather at this crucial juncture in our global journey toward sustainability, CEGS-2024 stands as a testament to our collective resolve and commitment. Together, we embark on a path toward a sustainable future, fueled by knowledge, innovation, and collaboration.

As Conference Chair, I have witnessed the power of collective effort and am continually inspired by the innovations and solutions that emerge when diverse minds collaborate. This conference and the proceedings within are a testament to that power. Join us on this transformative journey as we chart a course toward a sustainable future. Together, let's turn aspiration into tangible reality.

Vancouver, BC, Canada
December 2023

Danish Mir Sayed Shah

Research and Education Promotion Association (REPA) Scientific Committee

Keynote Speakers

Dr. Danish Mir Sayed Shah

Energy Systems (Chubu Electric Power) Funded Research Division, Nagoya University, Japan
Presentation Title: *Technological Advancements: Sustainable Energy and Green Innovations*

Dr. Siti Norliyana Harun

Research Fellow/ Senior Lecturer, Centre for Tropical Climate Change System, Institute of Climate Change, The National University of Malaysia, Malaysia
Presentation Title: *Transforming Waste into Wealth for a Greener Tomorrow: A Case Study of Rice Straw Valorization for Bioenergy Production*

Dr. Alexey Mikhaylov

Deputy Director of Monetary Relations Research Center, Financial University under the Government of the Russian Federation, Russia
Presentation Title: *Economic Aspects of Sustainability: Green Economy and Sustainable Business Practices*

Dr. Adnan Ahmed Sheikh

Associate Professor, Multan Campus, Air University Islamabad, Pakistan
Presentation Title: *How AI Enabled Blockchain Technology and Green Innovation Enhances Sustainable Business Performance*

Dr. Yogendra Narayan

Associate Professor, Chandigarh University, India
Presentation Title: *Hybrid Control of a Robotic Device Using Bio-Medical Signals*

Dr. Vivek Kumar Singh

Principal Consultant, Net Zero Think Pvt, India
Presentation Title: *India's Climate Promise: A Pathway to Net Zero Emissions via Energy Transition*

Contents

Data-Driven Pathways to Sustainable Energy Solutions

Mir Sayed Shah Danish, Mikaeel Ahmadi, Abdul Matin Ibrahimi, Hasan Dinçer, Zahra Shirmohammadi, Mahdi Khosravy, and Tomonobu Senjyu

1 Introduction

Machine learning, a subfield of AI, creates algorithms that learn from data. It encompasses supervised learning, which uses labeled data; unsupervised learning, which detects patterns; and reinforcement learning, which relies on feedback to reach goals [1].

These techniques enhance multicriteria decision-making in developing effective energy policies and strategies. Multicriteria analysis, often termed multicriteria decision-making, falls under operations research and specializes in decision-making issues. This type of problem involves selecting option(s) from multiple alternatives and is defined by the need to make a choice [2]. The decision-maker, who selects the

M. S. S. Danish (✉)
Energy Systems (Chubu Electric Power) Funded Research Division, Nagoya University, Nagoya, Japan
e-mail: danish@mdanish.me

M. Ahmadi · A. M. Ibrahimi
Research Promotion Unit, Co-Creation Management Department, University of the Ryukyus, Okinawa, Japan

H. Dinçer
School of Business, Istanbul Medipol University, Istanbul, Turkey

Z. Shirmohammadi
Faculty of Computer Engineering, Shahid Rajaee Teacher Training University, Tehran, Iran

M. Khosravy
Cross Laboratories, Cross-Compass Ltd., Tokyo, Japan

T. Senjyu
Department of Electrical and Electronic Engineering, University of the Ryukyus, Okinawa, Japan

M. S. S. Danish (ed.), *Unified Vision for a Sustainable Future*,
https://doi.org/10.1007/978-3-031-53574-1_1

preferred option(s), bases their decision on a set of criteria, typically more than two. The decision-maker's preferences are crucial when determining the best choice(s).

In the scope of multicriteria decision-making for energy policy and strategy development, sustainability and efficiency continue to be the primary influencing factors. Efficiency became a focal point by the late nineteenth century, concurrent with the advent of global industrialization and the onset of commercial energy generation and trade [3]. While essential criteria for achieving sustainability include accessibility, affordability, equity, safety, user efficiency, supply and production efficiency, cost-effectiveness, and the mitigation of environmental impacts on air, water, and soil quality [3].

In the energy sector, efficient data management becomes critical as it directly influences forecasting, decision-making, and sustainable energy distribution [4].

Energy policy encompasses the guidelines for shaping or maintaining energy systems'development, incorporating a cycle of goal setting, budgeting, execution, and monitoring [5]. A novel energy policy framework, highlighting the analysis of resource deployment impacts, the establishment of energy-related indicators and indices, the reconciliation of multidisciplinary decision-making indicators, and a comparative effectiveness analysis against the existing literature, is essential. Many developing countries are frequently having difficulty in developing dynamic policies in an emerging national and community levels' economic. Always these challenges associated with plenty of opportunities to be identified varying interests, and reconcile their different perspectives in national level.

2 Sustainability Attributes in the Energy Sector

The recent global trends in society modernization and lifestyle change alter an increasing demand for energy, while the production of environmentally friendly and cost-effective energy remains a continuing matter of interest within multidisciplinary [6]. As a matter of fact, renewable energy has drawn much attention nowadays because of its nexus to restrain global warming and support sustainable development goals (SDGs):

- Social Impact of Energy Production: Energy production from primary sources significantly impacts the quality of life in urban and rural areas. Women and children, especially in remote communities, often struggle to access clean and sustainable energy, which is critical for a healthy lifestyle [7].
- Economic and Social Interplay: A healthy lifestyle is closely tied to economic conditions, where residents' income determines their standard of living. In many developing countries, women typically have little or no income, limiting their participation in community decisions and access to healthier lifestyles [8]. Consequently, they often depend on primary energy sources.
- Diversity and Disparity in Energy Access: Addressing the disparities in energy access demands concerted efforts. Adopting inclusive approaches at the

community level can unveil opportunities that may be overlooked at broader national or regional levels [9]. A thorough community-level examination of energy demand and cultural practices, growth rates, economic and social drivers, and modernization trends could reveal alternatives to primary energy sources, leading to improved health outcomes, economic recovery, and long-term benefits across environmental, economic, technical, and social dimensions.

- Role of Energy Policy in Social Sustainability: Energy policies can play a crucial role in fostering social sustainability, particularly in marginalized communities. These policies can identify and support the integration of individuals into broader socioeconomic development, professionalizing their contributions and ensuring inclusive progress [10].
- Economic Benefits of Energy Policy in the Twenty-First Century: In today's economies, well-crafted energy policies can translate into economic benefits for all stakeholders, including end-users [11]. Distributing the economic gains from energy equitably can foster mutual trust between suppliers and consumers, enhancing reliability, efficiency, and overall sustainability.
- Reassessing Energy Policies: In practice, energy policies need regular reassessment to address potential inefficiencies and negative impacts associated with energy production and distribution [12]. This reevaluation should consider all aspects of the energy lifecycle, from production to consumption, to ensure that services meet realistic and sustainable standards.

3 Policy Relent Dilemma on Strategic Linkage

As a matter of fact, many developing countries are confronting a lack of exhaustive, viable policies to rationally overcome existing policy-related issues. These countries need a policy choice closely linked to a national strategy that results in an organized and viable set of actions [13]. One of the main drawbacks of these policies has been their independence from the parent strategy, which implies that policy actions need to be revised repeatedly or somehow be wholly inconsistent with national goals and objective priorities.

However, the starting point for policy development is the first step, and moreover, it plays a key role in the successful implementation of the policy. This step sets the high-level goals of a policy within a clear purpose of performance and priorities at the national level, backed by national and international legislation, rules, and standards. From an academic research standpoint, this phase is likewise a baseline study to evaluate necessities, assess possibilities, identify stakeholders' inferentiality, assemble resources, and define the development strategy. The analysis constituents of this step are the scope statement, business case analysis, milestone definitions, resource assessment, stakeholders' identification, responsibility distribution, and authorities' control [14].

4 Role of Machine Learning in the Energy Sector

The application of neural networks (NNs) in the energy sector has been transformative, enabling accurate demand forecasting, optimizing grid distributions, and predicting equipment failures [15]. Neural networks refer to a series of algorithms that aim to recognize the underlying relationships in a set of data, inspired by the way the human brain operates [16]. The human brain contains about 20 billion interconnected neurons that transmit and receive signals through electrochemical processes, converting electrical pulses into chemical messages for other neurons [17]. Dendrites receive signals, synapses track significant inputs, and the soma evaluates if inputs reach a threshold. The axon handles transmission. Despite human neurons having a slower switching speed (10^{-3} s) than computers (10^{-10} s), the brain can recognize a person in one-tenth of a second through parallel neuron operations. The neural network was introduced in 1943 as a mathematical model as the core component of deep learning and a subset of machine learning, drawing inspiration from the human brain's interconnected neurons to efficiently process and analyze complex data through parallel operations and adaptive learning capabilities [18].

As a starting point for understanding neural networks'potential in the energy sector, Table 1 shows the classification of neural networks based on the perceptron structure, functions, performance, and energy sector applications. Performance and applicability may vary with problem complexity, dataset size, and network architecture. The listed energy sector applications are nonexhaustive, and the accuracy depends on factors like hyperparameter tuning and data quality training. As neural network research progresses, performance and applicability in various sectors, including energy, will evolve.

The provision of a comprehensive machine learning tutorial, underpinned by extensive data, offers researchers a practical guide from beginner to advanced levels. Furthermore, the presented comparison table of various neural network types, focusing on their applications in the energy sector, provides an easy-to-use tool for researchers to select suitable models for specific tasks.

Figure 1 provides an exhaustive visual representation of the machine learning model development lifecycle, encapsulating the critical stages from initial data preparation to final model deployment. It emphasizes the interdisciplinary nature of AI, integrating domain expertise with statistical mathematics and computer science, and delineates the various learning paradigms—supervised, semisupervised, and reinforcement learning—that are fundamental to algorithm training and predictive accuracy.

Table 2 functions as an intricate roadmap for understanding the various components, techniques, and models intrinsic to the field of neural networks and their implementation across different network types. It meticulously categorizes each element, ranging from foundational layers such as the input, hidden, and output layers to sophisticated techniques like batch normalization and recurrent layers. For instance, it elucidates how input layers serve as the initial receptor of data, with examples including image pixels in a convolutional neural network (CNN), while

Table 1 Categorization of neural network types according to the perceptron architecture, primary functions, performance levels, and applications within the energy sector [19–21]

Neural network	Perceptron	Function	Performance	Most used applications in energy sector
Feedforward neural network (FNN)	Monolayer or multilayer	Classification, regression, and function approximation	Moderate, depending on the complexity of the problem and network architecture	Load forecasting, renewable energy generation prediction, energy management
Convolutional neural network (CNN)	Multilayer	Image classification, object detection, and image segmentation	High in image-related tasks, efficient in handling spatial data	Detection of solar panel defects, wind turbine fault diagnosis, satellite imagery analysis for site selection
Recurrent neural network (RNN)	Multilayer	Sequence prediction, time series analysis, and natural language processing	Moderate to high, depending on problem complexity and architecture	Short-term load forecasting, energy consumption prediction, equipment failure prediction
Long short-term memory (LSTM)	Multilayer	Sequence prediction, time series analysis, and natural language processing	High, especially in handling long-range dependencies in sequential data	Long-term load forecasting, renewable energy generation prediction, equipment failure prediction
Gated recurrent unit (GRU)	Multilayer	Sequence prediction, time series analysis, and natural language processing	Slightly faster than LSTM but with comparable performance	Short-term load forecasting, energy consumption prediction, equipment failure prediction
Radial basis function network (RBFN)	Monolayer or multilayer	Classification, regression, and function approximation	Moderate and efficient in handling smaller datasets and specific problems	Load forecasting, energy management, equipment fault diagnosis
Autoencoder (AE)	Multilayer	Dimensionality reduction, feature learning, and denoising	Moderate, depending on the complexity of the problem and network architecture	Feature extraction for fault detection, anomaly detection in energy data, energy data compression

(continued)

Table 1 (continued)

Neural network	Perceptron	Function	Performance	Most used applications in energy sector
Deep belief network (DBN)	Multilayer	Classification, regression, feature learning, and unsupervised pretraining	Moderate to high, depending on the complexity of the problem and network architecture	Load forecasting, renewable energy generation prediction, energy management
Generative adversarial network (GAN)	Multilayer	Generative modeling, image synthesis, and data augmentation	High, especially in generating realistic data samples	Data augmentation for energy-related tasks, synthetic energy data generation, optimizing energy systems

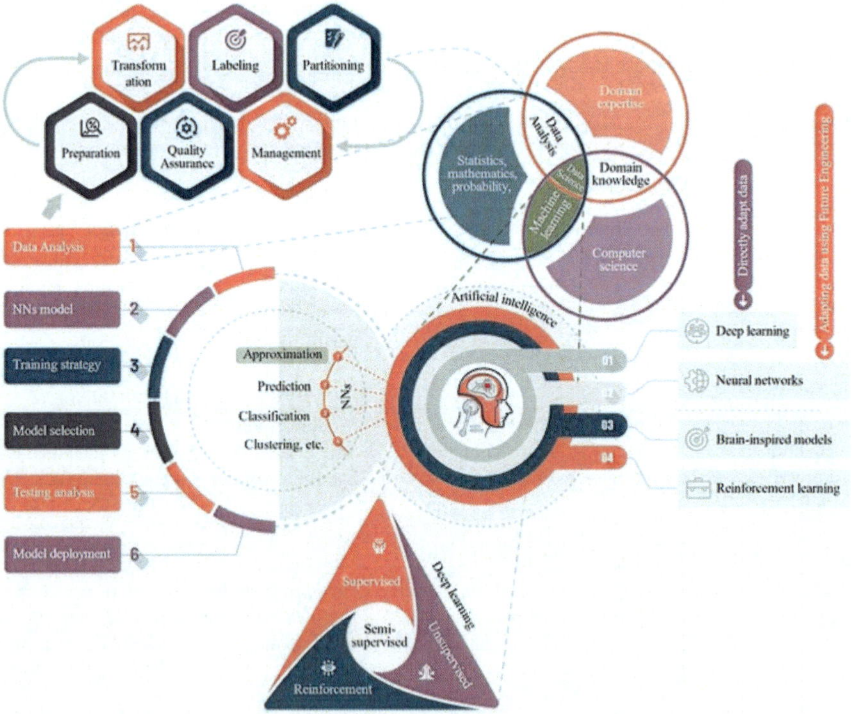

Fig. 1 Cutting-edge schema for computational and quantitative sciences

Table 2 Detailed representation of neural network building blocks [22–29]

No.	Category	Building block	Purpose	Functionality	Parameters	Example
1	Layers	Input layer	Receive input data	Provides the initial data to the neural network	–	Image pixels as input to a CNN
2	Layers	Hidden layer(s)	Perform computations	Processes and transforms the data through weights, biases, and activations	–	Multiple hidden layers in a deep NN
3	Layers	Output layer	Produce the final result	Generates predictions or classifications	–	Output layer in a binary classifier
4	Components	Neurons	Basic processing units	Receive input signals, process them, and generate output signals	–	Neurons in a fully connected layer
5	Components	Weights	Determine the influence of inputs on the output	Numerical values associated with the connections between neurons	W (weight matrix)	Weights in a fully connected layer
6	Components	Biases	Shift the decision boundary	Constants added to the weighted sum of inputs	b (bias vector)	Bias term in a fully connected layer
7	Components	Activation functions	Introduce nonlinearity	Transform the weighted sum of inputs into the output of a neuron	–	ReLU in a CNN hidden layer
8	Techniques	Initialization	Set initial values for weights and biases	Assign starting values to the weights and biases in the network	–	Glorot initialization
9	Techniques	Weight sharing	Reduce the number of parameters	Sharing weights across multiple neurons to reduce computational complexity	–	Weight sharing in convolutional layers
10	Techniques	Loss function	Quantify the network's performance	Measure the difference between predicted outputs and actual target values	–	Crossentropy loss for classification
11	Techniques	Optimizer	Adjust the weights and biases	Algorithm used to minimize the loss function	–	Adam optimizer
12	Techniques	Learning rate	Control the update step in optimization	Determine the step size during optimization	α (alpha)	Learning rate of 0.001 in an optimizer

(continued)

Table 2 (continued)

No.	Category	Building block	Purpose	Functionality	Parameters	Example
13	Techniques	Momentum	Accelerate convergence and avoid local minima	Add a fraction of the previous update to the current update	β (beta)	Momentum of 0.9 in an optimizer
14	Techniques	Hyperparameters	Control model complexity and training process	Tunable parameters that dictate the model's architecture and learning dynamics	–	Number of hidden layers, learning rate, dropout
15	Techniques	Regularization	Prevent overfitting	Penalize complex models or constrain weights to prevent overfitting	λ (lambda)	L2 regularization (ridge regression)
16	Techniques	Dropout	Reduce overfitting	Randomly drop neurons during training to prevent overreliance on any one neuron	Dropout rate	Dropout rate of 0.5 in a deep learning model
17	Techniques	Normalization	Improve training stability and convergence speed	Scale input features to have similar distributions	–	Standardization of input data
18	Techniques	Batch normalization	Accelerate training and improve generalization	Normalize activations of each layer before the activation function	–	Batch normalization in a CNN
19	Techniques	Gradient clipping	Prevent exploding gradients	Limit the size of the gradients during backpropagation	Clipping threshold	Gradient clipping in an RNN
20	Techniques	Backpropagation	Update weights and biases	Compute gradients of the loss with respect to weights and biases	–	Backpropagation in a feedforward neural network
21	Techniques	Epochs	Control training iterations	Number of times the entire training dataset is passed through the network	–	Training for 100 epochs
22	Techniques	Mini-batches	Speed up training and improve convergence	Divide the training dataset into smaller subsets for faster updates	Mini-batch size	Mini-batch size of 32
23	Techniques	Validation set	Evaluate model performance during training	A subset of the dataset used to evaluate the model and prevent overfitting	–	80/20 train/validation split

24	Techniques	Metrics	Measure model performance	Quantify the quality of the model's predictions	–	Accuracy metric for classification
25	Techniques	Learning rate scheduling	Adjust learning rate during training	Modify the learning rate based on a schedule or the current state of training	–	Reduce learning rate on plateau
26	Techniques	Early stopping	Prevent overfitting	Stop training when the validation performance no longer improves	Patience, metric	Early stopping with patience of 10 epochs
27	Layers	Transfer learning	Leverage pretrained models	Fine-tune a pretrained model for a new task with a smaller dataset	–	Fine-tuning a pretrained ResNet on a new dataset
28	Layers	Pooling layers	Reduce spatial dimensions	Downsample the feature maps by aggregating local spatial information	–	Max pooling in a CNN
29	Layers	Convolutional layers	Learn spatial features	Apply convolution operations to learn local spatial features	Filter size, stride, padding	Convolutional layer in a CNN
30	Techniques	Recurrent layers	Model temporal dependencies	Maintain hidden states to learn temporal patterns in sequential data	–	LSTM layer in an RNN
31	Techniques	Skip connections	Improve gradient flow	Add shortcuts between layers to allow gradients to bypass some layers	–	Residual connections in a ResNet
32	Techniques	Attention mechanisms	Learn to focus on relevant information	Weight the input features based on their relevance to the output	–	Self-attention in a transformer
33	Models	Embeddings	Convert categorical data into continuous vectors	Map discrete categories into continuous space	Embedding size	Word embeddings in an NLP model
34	Models	Autoencoders	Learn efficient representations	Unsupervised learning to compress and reconstruct input data	–	Denoising autoencoder
35	Techniques	Generative models	Generate new data	Learn the underlying data distribution to create new samples	–	Generating images with a GAN

(continued)

Table 2 (continued)

No.	Category	Building block	Purpose	Functionality	Parameters	Example
36	Techniques	Adversarial training	Improve robustness	Train a model with adversarial examples to improve its generalization	–	Adversarial training for a classifier
37	Techniques	Test set	Evaluate final model performance	A held-out dataset used to assess the model's performance on unseen data	–	70/15/15 train/validation/test split
38	Techniques	Model selection and hyperparameter tuning	Optimize model configuration	Search for the best combination of hyperparameters and model architecture	–	Selecting the best model from a set of candidates
39	Techniques	Grid search	Exhaustive hyperparameter search	Search over a predefined set of hyperparameter values	Grid search space	Grid search for SVM parameters
40	Techniques	Random search	Efficient hyperparameter search	Sample hyperparameter values from a distribution	Search space	Random search for deep learning model parameters
41	Techniques	Bayesian optimization	Guided hyperparameter search	Model the relationship between hyperparameters and performance to guide search	Acquisition function	Bayesian optimization for a neural network
42	Techniques	Model predictions	Generate output for new data	Apply the trained model to make predictions on unseen data	–	Predictions on test data using a trained model
43	Techniques	Confidence scores	Quantify uncertainty of predictions	Provide a measure of the model's certainty in its predictions	–	Softmax output for classification probabilities
44	Techniques	Feature importance	Identify influential input features	Rank input features based on their contribution to the model's predictions	–	Permutation importance in a random forest
45	Report	Visualization of learned features	Interpret model's internal representations	Visualize features learned by the model to gain insights into its decision-making	–	Visualizing filters in a CNN

1	–	Hidden layer(s), output layer	All types	Directly receive input data	Limited to input data size
2	–	Input layer, output layer, neurons, weights, biases, activation functions	All types	Enable learning of complex patterns	May require extensive training
3	–	Input layer, hidden layer(s), neurons, weights, biases, activation functions	All types	Generate final predictions	May require fine-tuning
4	–	Layers, weights, biases, activation functions	All types	Core processing units of NNs	May require fine-tuning
5	–	Neurons, activation functions, initialization	All types	Determine the influence of inputs	May require tuning
6	–	Neurons, activation functions, initialization	All types	Shift the decision boundary	May require tuning
7	Varies	Neurons, weights, biases	All types	Introduce nonlinearity	Choice depends on the problem
8	Varies	Weights, biases	All types	Set initial values for weights and biases	Choice depends on the problem
9	Limited flexibility	Convolutional layers	CNNs	Reduce the number of parameters	May limit model flexibility
10	Varies	Optimizer, metrics	All types	Quantify network performance	Choice depends on the problem
11	Varies	Loss function, learning rate, momentum, hyperparameters	All types	Adjust weights and biases	Choice depends on the problem
12	Sensitivity to tuning	Optimizer, momentum, learning rate scheduling	All types	Control the step size in optimization	Sensitive to tuning
13	Sensitivity to tuning	Optimizer, learning rate	All types	Accelerate convergence, avoid local minima	Sensitive to tuning
14	Sensitivity to tuning	Dropout, normalization, batch normalization	All types	Prevent overfitting	Choice depends on the problem

(continued)

Table 2 (continued)

No.	Limitations	Related building blocks	Applicable network types	Advantages	Disadvantages
15	Impact on model complexity	Regularization	All types	Reduce overfitting by randomly dropping neurons	May increase training time
16	Reduced model capacity	Regularization, batch normalization	All types	Scale input features to a standard range	May require additional processing
17	Sensitive to outliers	Regularization, normalization	All types	Normalize layer inputs across minibatches	Sensitive to batch size
18	Computationally expensive	Backpropagation	All types	Limit gradient magnitude to prevent exploding gradients	Choice of threshold matters
19	Arbitrary clipping value	Epochs, mini-batches, gradient clipping	All types	Compute gradients for weight updates	Prone to vanishing/exploding gradients
20	Computationally expensive	Minibatches, validation set, early stopping	All types	Determine the number of full passes through the dataset	Choice depends on the problem
21	Sensitivity to tuning	Backpropagation, epochs, validation set	All types	Speed up training and reduce memory requirements	Choice of batch size matters
22	Sensitivity to tuning	Epochs, minibatches, metrics, early stopping	All types	Estimate model performance during training	Reduces training data size
23	Dataset size limitations	Loss function, validation set, test set	All types	Evaluate model performance	Choice depends on the problem
24	Not always comprehensive	Learning rate, optimizer	All types	Adjust learning rate during training	Requires tuning
25	Sensitivity to tuning	Validation set, epochs, metrics	All types	Prevent overfitting by stopping training when validation performance no longer improves	May require a validation set and tuning
26	May stop too early or too late	Pretrained models	All types	Use pretrained models to speed up training and improve performance	Requires compatible architecture and dataset
27	Limited to similar tasks	Convolutional layers	CNNs	Reduce spatial dimensions and control overfitting	May lose some information

28	Loss of spatial information	Pooling layers, weight sharing	CNNs	Learn spatial hierarchies and local patterns	Specific to image and grid-like data
29	Assumes spatial structure	Attention mechanisms	RNNs, LSTMs, GRUs	Learn temporal dependencies and sequences	Prone to vanishing/exploding gradients
30	Gradient vanishing/ exploding	Convolutional layers, recurrent layers	ResNets, DenseNets	Improve gradient flow and mitigate vanishing gradients	May increase model complexity
31	Computationally expensive	Recurrent layers, embeddings	Transformers, RNNs, LSTMs	Focus on important parts of the input	May increase model complexity
32	Computationally expensive	Attention mechanisms	All types, specific to NLP	Map categorical data to dense vectors	Require sufficient data to learn embeddings
33	High-dimensional space	Encoder, decoder	Unsupervised learning	Learn representations and compress data	May lose information during compression
34	Limited reconstruction quality	Variational autoencoders, GANs	Unsupervised learning	Generate new samples from the learned distribution	May require extensive training
35	Mode collapse, instability	Generative models	GANs	Improve model robustness and generalization	May be difficult to train
36	Increased training complexity	Metrics, validation set	All types	Evaluate model performance on unseen data	Reduce training data size
37	Dataset size limitations	Grid search, random search, Bayesian optimization	All types	Find the best model and hyperparameter values	May require extensive search
38	Computationally expensive	Model selection, random search, Bayesian optimization	All types	Exhaustive hyperparameter search	Computationally expensive

(continued)

Table 2 (continued)

No.	Limitations	Related building blocks	Applicable network types	Advantages	Disadvantages
39	Computationally expensive	Model selection, grid search, Bayesian optimization	All types	Randomized hyperparameter search	Less exhaustive and may miss optimal values
40	Suboptimal configurations	Model selection, grid search, random search	All types	Guided hyperparameter search	More complex and may require tuning
41	Expensive model evaluations	Test set, metrics	All types	Generate predictions from the trained model	Depends on the quality of the model
42	Dependent on model quality	Model predictions	All types	Estimate prediction confidence	Depends on the quality of the model
43	May not reflect true uncertainty	Model predictions	All types, specific to tree-based models	Rank input features by importance	Depends on the model and data
44	Dependent on model quality	Model predictions, autoencoders	All types	Visualize learned representations and features	Requires suitable visualization techniques
45	May not be interpretable	Related building blocks	Applicable network types	Advantages	Disadvantages

hidden layers delve into the computations through activations and weights, indicative of a deep neural network's multiple hidden layer architecture. Techniques like weight sharing and dropout are highlighted for their roles in reducing parameters and preventing overfitting, respectively. The table also delves into loss functions, optimizers, and learning rates, which are pivotal in gauging and refining network performance. Additionally, it touches upon more subtle concepts like attention mechanisms, which allow models to focus on relevant parts of the input, and generative models, which are capable of producing new diverse data samples. Limitations are candidly addressed for each component, such as the sensitivity of activation functions to the problem at hand or the increased computational demand of techniques like backpropagation. Furthermore, the table delineates the interconnectedness of these elements, for example, dropout related to regularization techniques and recurrent layers linked to temporal dependency modeling. Advantages such as the capability of pooling layers to learn spatial hierarchies are balanced against disadvantages like potential information loss. This analytical categorization guides the understanding of not only each building block's function, advantages, and limitations but also their interrelationships and applicability to various network types, offering a holistic overview that is essential for both beginners and seasoned professionals in the field of neural networks.

Selected applications of machine learning, particularly neural networks, for the observation and optimization of energy systems are shown in Table 3. The table presents a comprehensive overview of the applications of neural networks in the enhancement of energy system monitoring and efficiency improvements. It highlights a range of models and methodologies applied in the solar and wind energy sectors from 2013 to 2023. For solar power, the methods vary from basic neural networks (NNs), which achieve a normalized mean absolute error (nMAE) of 7.5% in solar radiation prediction, to advanced models such as long short-term memory (LSTM) networks, which offer day-ahead sun radiation forecasting with a root mean square error (RMSE) of 18.34%. Remarkably, hybrid models that combine recurrent neural networks and shallow neural networks demonstrate superior precision, with an RMSE as low as 0.19%. In the realm of wind power, generative adversarial networks (GANs) are utilized for real-time forecasting and multilayer perceptrons (MLP) for predicting wind speeds on turbine surfaces with a mean absolute percentage error (MAPE) of 1.479%. The integration of autoregressive integrated moving average (ARIMA) with artificial neural networks (ANNs) in tracking the Internet of Things (IoT)-powered system indicates a significant potential for boosting energy production efficiency despite a MAPE of 32.0%. This table also sheds light on the progression of neural network complexity over the years and their varied accuracy rates assessed using standard metrics like RMSE, MAPE, and nMAE, offering essential quantitative insights for future innovations in the field of energy system analytics.

Table 3 Applications of neural networks in energy systems observation and optimization

System	Model/ method	Year	Description	Accuracy measure	References
Solar power	NNs	2023	Providing fast and accurate solar radiation predictions based on limited observation data	nMAE 7.5%	Jia et al. [30]
Solar power	NNs	2023	Offering an understanding of recursive feature elimination and its integration with different models in multivariate solar radiation forecasting	RMSE 0.3%	Hissou et al. [31]
Solar power	NNs	2023	Forecasting models based on a hybrid architecture that combines recurrent neural networks and shallow neural networks	RMSE 0.19%	Castillo-Rojas et al. [32]
Wind power	GANs	2022	Generative adversarial networks for real-time wind power forecasting	Various	Bentsen et al. [33]
Wind power	MLP	2022	Measuring and predicting wind speed on the wind turbine surface	MAPE 1.479%	Zhang et al. [9]
Solar power	ARIMA-ANN	2021	Using a tracking IoT-powered system to improve energy production efficiency	MAPE 32.0%	Adli et al. [34]
Wind power	FFBP-ANN	2020	Atmospheric parameters impact evaluation on wind power curve	nMAE 59%	Nielson et al. [35]
Wind power	FFBP-ANN, RBF-ANN	2019	Time series input influence the NN performance along with learning rate changes	MAE = 1.112 m/s	Chen et al. [36]
Solar power	LSTM	2018	Day ahead forecasting of sun radiation using meteorological information	RMSE 18.34%	Qing and Niu [37]
Solar power	ANN and fuzzy logic	2017	For forecasting solar power, a triple layer BP, fuzzy preprocessing, and an ANN were combined	MAPE 29.6%	Sivaneasan et al. [38]
Solar power	ANN	2016	Forecasting global solar irradiance using different atmospheric and environmental factors	nRMSE 20%	Gutierrez-Corea et al. [39]
Solar power	FF-ANN	2015	Preprocessing of input data, clustering, and elimination of night hours to enhance the accuracy of predictions	R-squared 96.65%	Abuella and Chowdhury [40]
Solar power	ANN-MLP	2014	Evaluating radiation data, including horizontal extraterrestrial irradiance, solar declination, and zenith angle.	RMSE 8.81%	Dahmani et al. [41]
Solar power	ANN	2013	Radiation forecasting model using velocimetry, cloud indexing, and solar irradiation	RMSE 5–25%	Marquez et al. [42]
Wind power	SVM	2012	SVM models solve the issue of local optimality in other machine learning networks	nMAE 54%	Tabari et al. [43]

5 Critical Analysis of Neural Network Prerequisites for Implementation in the Energy Sector

5.1 Data

Data is a critical component in machine learning and analytics, undergoing multiple processing stages to ensure quality and usability, yet the term big data remains conceptually vague despite its popularity in academia and industry [44, 45]:

- Data collection is the process of gathering raw data from various sources for analysis and processing. Applications of this process include but are not limited to utility company data, stock exchange trends data, market research, customer feedback, and IoT sensor data. Data can be collected through web scraping, application programming interfaces (APIs), surveys, databases, Internet of Things (IoT) devices, and manual input. Data collection provides the foundation for analysis and modeling, while limitations include the potential for incomplete, biased, or irrelevant data. Ensuring a diverse and representative sample is essential for reliable results.
- Data cleansing involves identifying and correcting errors, inconsistencies, and inaccuracies in datasets, using it to improve data quality, ensure accurate analysis, and improve model performance. The process includes removing duplicates, fixing typos, correcting inconsistencies, and handling missing values. Data cleansing offers advantages such as increased data reliability and reduced noise in the dataset. However, it can be time-consuming and may introduce errors if not performed carefully. Automation and manual review can help maintain data quality.
- Data labeling assigns meaningful tags, labels, or annotations to data points. It is crucial for supervised learning, where models need labeled data to learn patterns. Methods for data labeling include manual annotation, crowdsourcing, semisupervised learning, and active learning. Data labeling improves model performance by providing ground truth for training and evaluation. It demonstrates, with some limitations, the potential for human error, subjectivity, and cost. Ensuring consistency and quality in the labeling process is critical.
- Data augmentation is the technique to increase the diversity and size of a dataset by creating new data points through transformations. This is particularly useful in computer vision, natural language processing, and audio processing. Data augmentation techniques include image rotations, flipping, cropping, scaling, noise injection, and time stretching for audio. It can improve model generalization, increase dataset size, and reduce overfitting. This technique sometimes shows potential distortion or loss of information and increased training time. Selecting the appropriate enhancement methods is highly recommended.
- During the data encoding process, raw data are converted into a format suitable for machine learning models. It is important for categorical data and text data representation. Methods include one-hot encoding, label encoding, binary encod-

ing, and word embeddings for text. Proper data encoding enables models to learn patterns and relationships in the data efficiently. However, some encoding methods can lead to increased dimensionality, sparsity, or loss of information. Choosing the right encoding method depends on the problem and the data type.

- Feature extraction is identifying and extracting relevant features from raw data to reduce dimensionality and improve model performance. Image recognition, speech recognition, and text classification are the main applications of this technique. Principal component analysis (PCA), linear discriminant analysis (LDA), autoencoders, and t-distributed stochastic neighbor embedding (t-SNE) with the merit of reduced computational complexity, improved model performance, and noise reduction are the primary techniques of this process. Limitations include the potential loss of information and interpretability. Careful selection of feature extraction techniques is necessary.
- The process of feature scaling is the transformation of numerical features to a common scale, ensuring that no feature dominates the model due to differences in magnitude. This technique applies in the context of gradient-based optimization and distance-based algorithms. Scaling methods include min-max scaling, standardization, and Mean normalization. Feature scaling improves the convergence speed and model performance. However, it may not be suitable for all datasets or algorithms and can sometimes reduce interpretability. Understanding the underlying data distribution and algorithm requirements is essential.
- Feature engineering involves creating new features from existing data to improve model performance and interpretability. This process applies to predictive modeling and pattern recognition, using polynomial features, interaction terms, and domain-specific transformation methods. Feature engineering can lead to improved model performance and insights. However, it can be time-consuming, require domain expertise, and increase model complexity. Expert knowledge and iterative experimentation are crucial in effective feature engineering.
- Data imputation is the process of replacing missing or incomplete data with estimated values to maintain data integrity and handling missing data in surveys, sensor data, and time series. Imputation methods include mean, median, mode, and K-nearest neighbors (KNN) imputation. Data imputation helps maintain data consistency, reduces bias, and improves model performance. It primarily constrains the potential introduction of noise or inaccurate estimations. Selecting the appropriate imputation method based on data distribution and domain knowledge is crucial.
- The data integration process combines data from multiple sources into a unified, coherent dataset, which applies to merging datasets for holistic analysis, customer data consolidation, and sensor fusion by joining tables, concatenation, and data fusion. Data integration provides a comprehensive view, leading to better insights and decision-making [2]. Limitations include potential data inconsistencies, privacy concerns, and increased complexity. Ensuring data compatibility and consistency is essential for successful data integration.
- Dimensionality reduction is the process of reducing the number of features in a dataset while retaining as much relevant information as possible. Dimensionality

reduction can lead to faster training, reduced overfitting, and improved interpretability. Limitations include the potential loss of information and reduced model performance. Selecting the appropriate dimensionality reduction technique is based on the problem and the characteristics of the data.

- Anonymizing data involves removing personally identifiable information (PII) from data to protect privacy while maintaining data utility. It is used to share data for research, comply with data protection regulations, and maintain customer privacy, using data masking, generalization, and k-anonymity. Data anonymization helps preserve privacy and meet regulatory requirements. However, it can introduce challenges in data quality, utility, and potential re-identification risks. Balancing data utility and privacy protection is key in data anonymization.
- Data splitting process dividing a dataset into separate subsets for training, validation, and testing machine learning models through model evaluation, hyperparameter tuning, and preventing overfitting. Methods include train-test split, K-fold cross-validation, stratified sampling, and leave-one-out cross-validation. Data splitting helps to assess model performance and generalization capabilities. The primary constraint for this technique is the potential of overfitting or underfitting if the split is not representative. Ensuring a diverse and representative sample in each subset is decisive for reliable results.
- Data shuffling is randomly reordering data points in a dataset to ensure uniform distribution and avoid biases. Applying breaking patterns in the data, ensuring model generalization, and improving training efficiency by hiring randomization, stratified shuffling, and time-based shuffling are the main contexts of this process. Data shuffle helps to improve model performance and reduce overfitting. However, it can introduce challenges in time-series data, where temporal order matters. Understanding the data structure and problem requirements is essential when applying data shuffle.
- Data versioning is the practice of tracking changes to a dataset over time, allowing for easy rollback and collaboration. This method is used for managing data updates, auditing, and reproducibility. This method includes version control systems, incremental backups, and metadata tracking. Data versioning offers increased collaboration, traceability, and efficient experimentation. However, it may introduce challenges in storage and management complexity. A robust data versioning system is essential for large-scale projects and collaborative environments.
- Data storage maintains and organizes data to allow efficient access, retrieval, and analysis. This method can be applied in various ways, including long-term preservation, backup, and sharing of datasets. Options for data storage include cloud storage, local storage, and databases. Data storage ensures data availability, security, and integrity. However, it is associated with potential data loss, storage costs, and privacy concerns. The selection of the appropriate storage solution depends on factors such as data size, access requirements, and security considerations.
- Data validation is the process of evaluating data for correctness, completeness, and consistency, which are used for quality assurance, anomaly detection, and

model evaluation. These techniques include statistical tests, visualizations, and outlier detection. Data validation enhances data quality, identifies errors, and ensures accurate analysis. Among the critical limitations, the main points are the potential for false positives, subjectivity, and time-consuming manual validation. A combination of automated and manual validation methods can facilitate the maintenance of high data quality.

- Data monitoring is the ongoing real-time tracking and evaluation of dataset's quality, performance, and usage for detecting anomalies, ensuring data quality, and monitoring model performance. Performance metrics, real-time monitoring, and alerts and notifications are examples of this process that enables proactive problem resolution, improves data quality, and enhances model performance. The shortcomings are the potential for false alarms, monitoring overhead, and privacy concerns. A comprehensive data monitoring system maintains data quality and ensures reliable model performance.
- Having established the importance and intricacies of dataset management, it becomes imperative to understand how these datasets are processed. This leads us to study the role of optimizers, which harness the power of the data to train and refine neural network models.

5.2 Dataset

A dataset is a collection of structured or unstructured data points utilized in various fields, such as machine learning and neural networks (NNs), for analysis, training, and evaluation. There are three main types of datasets: structured, which consists of tabular data with a fixed schema like CSV and Excel files; unstructured, which lacks a predefined schema and includes text, images, and audio; and semistructured, a combination of structured and unstructured elements found in formats like XML and JSON. The structure of datasets can be tabular, hierarchical, or network-based. As in spreadsheets, tabular structures arrange data in rows and columns with labeled attributes. Hierarchical structures organize data trees like file systems and network structures connect data points through relationships, as seen in social networks and graphs. Various formats are used to store and represent datasets, including text formats such as CSV, TSV, and TXT; binary formats (HDF5, protocol buffers, and Parquet); markup formats (JSON, XML, and YAML); image formats (JPEG, PNG, and TIFF); and audio formats (WAV, MP3, and FLAC). Converting methods for datasets include parsing, which converts data from one format to another; serialization, the process of converting data objects into binary, text, or markup formats; deserialization, which converts data back to data objects from binary, text, or markup formats; encoding, transforming data into a machine-readable format; and decoding, converting data from a machine-readable format back to their original forms.

Dataset classification organizes various data processing tasks into six main categories, highlighting the different stages and aspects of handling data (Fig. 2) [44,

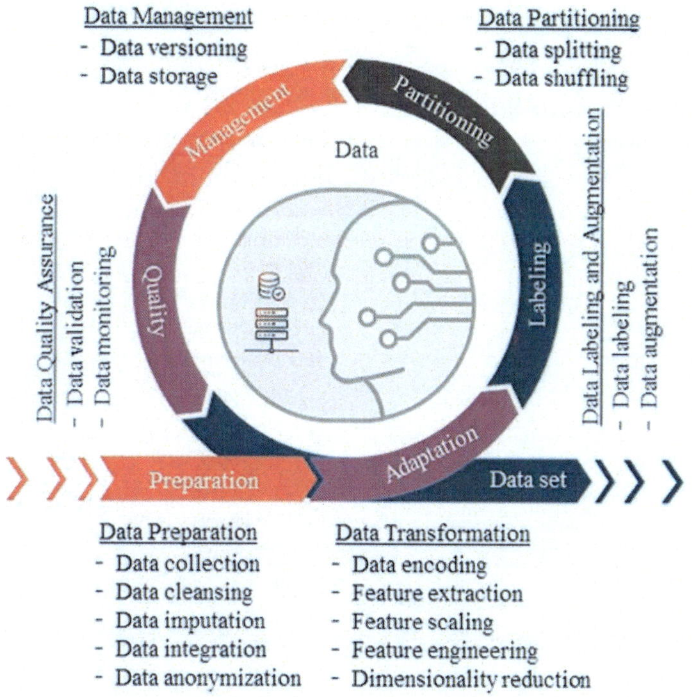

Fig. 2 Classification of six main categories of data processing tasks, emphasizing the various stages and aspects of handling data

46–48]. Data preparation involves collecting, cleaning, and integrating raw data to form a unified, high-quality dataset while also handling missing values and protecting sensitive information. Data transformation focuses on converting data into suitable formats, extracting important features, scaling, and engineering new features to enhance the dataset's effectiveness for machine learning or analytics tasks, including dimensionality reduction. Data labeling and augmentation involve annotating data for supervised learning tasks and augmenting the dataset with new or modified instances to improve model performance. Data partitioning consists of dividing the dataset into subsets for model evaluation, preventing overfitting, and ensuring unbiased training through data shuffling. Data management tracks changes, maintains historical records, and stores data in various forms for easy access and collaboration. Lastly, data quality assurance aims to ensure data correctness, consistency, and integrity through validation, visualization, outlier detection, and continuous monitoring of the data pipeline.

Data quality, ensuring accuracy, completeness, and consistency; data privacy, protecting sensitive information through anonymization or pseudonymization; data provenance, tracking the origin, transformation, and usage history of datasets; and data governance, establishing policies and processes for managing datasets throughout their lifecycle (Table 4).

Table 4 Comprehensive data preparation, transformation, and management techniques for machine learning and analytics

Class	Technique
Class 1: Data preprocessing techniques	
Data collection	Web scraping
	APIs
	Surveys
	Databases
	IoT devices
	Manual input
Data cleansing	Remove duplicates
	Fix typos
	Correct inconsistencies
	Handle missing values
Data labeling	Manual annotation
	Crowdsourcing
	Semisupervised learning
	Active learning
Data augmentation	Image rotations
	Flipping
	Cropping
	Scaling
	Noise injection
	Time stretching (audio)
Class 2: Data transformation techniques	
Data encoding	One-hot encoding
	Label encoding
	Binary encoding
	Word embeddings (text)
	Feature extraction
	Principal component analysis (PCA)
Linear discriminant analysis (LDA)	Autoencoders
	t-distributed stochastic neighbor embedding (t-SNE)
	Feature scaling
	Min-max scaling
Standardization	Mean normalization
	Feature engineering
	Polynomial features
	Interaction terms

(continued)

Table 4 (continued)

Class	Technique
Domain-specific transformations	Data imputation
	Mean imputation
	Median imputation
	Mode imputation
	K-nearest neighbors (KNN) imputation
	Data integration
	Joining tables
Concatenation	Data fusion
	Dimensionality reduction
	PCA
	LDA
Autoencoders	t-SNE
	Data anonymization
	Data masking
	Generalization
k-anonymity	Data splitting
	Train-test split
	K-fold cross validation
Stratified sampling	Leave-one-out cross validation
	Data shuffling
	Randomization
Stratified shuffling	Time-based shuffling
	Data versioning
	Version control systems
	Incremental backups
Metadata tracking	Data storage
	Cloud storage
	Local storage
Databases	Data validation
	Statistical tests
	Visualizations
	Outlier detection
Data monitoring	Performance metrics
	Real-time monitoring
	Alerts and notifications

5.3 Optimizers

Optimizers are used to efficiently minimize the loss function by adjusting the model parameters (weights and biases) to achieve optimal model performance [49]. A detailed overview of complexity, computation time, accuracy, advantages, disadvantages, formula, and formula abbreviations for a selection of widely used optimization algorithms in NNs intended to assist in selecting the most suitable optimizer for the specific tasks and problems is provided in Table 5. The optimizer matrix provides an organized and concise summary of various optimizers and their specific characteristics.

Comparative analysis of various optimizers can efficiently pinpoint the most fitting one for a specific model and problem, conserving both time and resources. The optimizer matrix enhances the comprehension of the strengths and weaknesses of different optimizers, guiding their application toward improved model performance. The matrix shown in Table 5 facilitates swift comparison and selection of optimizers for diverse models and scenarios.

The complexity, computation time, and accuracy of the rest of the most common optimizers, which are not listed in Table 5, are reported to be high, with the advantages of global optimization robust to local minima, and these optimizers are as follows: genetic algorithms, differential evolution, particle swarm optimization, Bayesian optimization, simulated annealing, cuckoo search, ant colony optimization, evolution strategies, natural evolution strategies, and covariance matrix adaptation evolution strategy (CMA-ES).

All of these optimizers are not employed as primary optimizers in NNs due to their high computational expense and slower convergence. However, these methods are included to offer a comprehensive overview of optimization techniques. It is important to note that the rankings in the table might not accurately represent each optimizer's actual complexity, computation time, and accuracy in all circumstances.

An optimizer's complexity, computation time, and accuracy can be highly dependent on the problem at hand, the specific implementation, and the chosen parameter settings. To identify the most suitable optimizer for a given problem, it is generally advised to experiment with various optimization techniques and compare their performance according to the task's specific requirements.

Recent advancements, especially in the energy sector, are leaning towards hybrid optimization techniques, which combine the strengths of gradient-based and evolutionary algorithms to achieve quicker and more accurate results [58]. While data collection methods, such as IoT devices, provide real-time data, challenges like data inconsistencies, missing values, and the sheer volume can pose hurdles. Similarly, while optimizers, such as SGD, can offer fast convergence, they may get trapped in local minima, affecting the overall NN performance [29].

Table 5 Classification of the most commonly used neural network optimizers [19, 26, 49–57]

	Optimizers	Nonlinear autoregressive (NAR)	Levenberg–Marquardt (LTK)	Divided symmetric search (DSS)	Cross-entropy function (CEF)	Generalized extreme value (GEV)	Linear matrix inequality (LMI)	Momentum (MOM)	Learning rate decay (LRD)	Adaptive learning rate (ALR)	Regularization (REG)
Group	Network architecture	Learning task	Dataset size	Computational efficiency	Gradient explosion or vanishing	Local minima	Momentum	Learning rate decay	Adaptivity (refers to adaptive learning rate methods)	Regularization (refers to techniques for preventing overfitting)	Network architecture
Gradient descent-based optimizers	Gradient descent (GD)	X	X	X	X	X	X				
	Stochastic gradient descent (SGD)	X	X	X	X	X	X				
	Minibatch gradient descent (MBGD)	X	X	X	X	X	X				
Momentum-based optimizers Adaptive learning rate optimizers	Momentum (MOM)	X	X	X	X	X	X	X			
	Nesterov accelerated gradient (NAG)	X	X	X	X	X	X	X			
	Adaptive gradient (ADG)	X	X	X	X	X	X		X	X	
	Adaptive delta (ADD)	X	X	X	X	X	X		X	X	
	Root mean square propagation (RMSProp)	X	X	X	X	X	X		X	X	
	Adam (ADM)	X	X	X	X	X	X	X	X	X	
	NAdam (NDM)	X	X	X	X	X	X	X	X	X	
	AdaMax (AMAX)	X	X	X	X	X	X	X	X	X	
	Adamax (AMD)	X	X	X	X	X	X		X	X	

(continued)

Table 5 (continued)

Group	Optimizers	Nonlinear autoregressive (NAR)	Levenberg–Marquardt (LTK)	Divided symmetric search (DSS)	Cross-entropy function (CEF)	Generalized extreme value (GEV)	Linear matrix inequality (LMI)	Momentum (MOM)	Learning rate decay (LRD)	Adaptive learning rate (ALR)	Regularization (REG)
	Network architecture	Learning task	Dataset size	Computational efficiency	Gradient explosion or vanishing	Local minima	Momentum	Learning rate decay	Adaptivity (refers to adaptive learning rate methods)	Regularization (refers to techniques for preventing overfitting)	Network architecture
Averaging and approximation-based optimizers	AMSGrad (AGD)	X	X	X	X	X	X				
	K-Fac (KFC)	X	X	X	X	X	X				
	Local Bayesian optimization (LBO)	X	X	X	X	X	X				
	Local-bias backpropagation (LBB)	X	X	X	X	X	X				
Gradient-free and proximal optimizers	Parameter reduction (PRD)	X	X	X	X	X	X				X
	Conjugate gradient descent (CGR)	X	X	X	X	X	X				
Regularization and adaptivity	Regularization (REG)	X	X	X	X	X	X			X	X

5.4 Hyperparameter Set

After selecting an optimizer, a hyperparameter set is required to be defined. For a specific optimizer, it can vary depending on the specific problem and dataset, and it may require experimentation and tuning to find the optimal combination of hyperparameters for a proposed optimization [29]. For instance, the use of time-series data in predicting electricity consumption patterns in a city or the implementation of the Adam optimizer in enhancing the wind energy production efficiency in off-shore farms [59].

Choosing the right hyperparameters is not just a theoretical exercise. In practical scenarios, like energy demand forecasting, slight tweaks in hyperparameters can mean the difference between accurate predictions, leading to efficient energy distribution, and massive energy wastages or outages.

6 Discussion

This study undertakes a comprehensive exploration of data management and neural network optimization, particularly emphasizing their roles in decision-making within the energy sector. The proposed methodical analysis reveals several significant findings. Traditionally, the energy sector has been driven by empirical and heuristic methods. This chapter underscores the transformative potential of structured data handling. By describing the differences between structured, unstructured, and semistructured datasets and elucidating their potential in energy decisions, we chart a new paradigm for energy-centric data analytics. While optimizers have been extensively studied in the context of general machine learning, their application in energy-specific neural network models remains sparse. It pioneers an exhaustive analysis of these optimizers, offering a unique perspective on their relevance to energy-centric problems.

The hyperparameter set guidance is tailored for energy datasets, making it a novel contribution. By understanding these hyperparameters in the context of energy, we can better optimize models for tasks such as load forecasting, renewable integration, and grid optimization.

The six main categories of data processing tasks, as highlighted in this chapter, offer a robust framework. Their application in the energy sector can revolutionize how data are perceived and utilized, shifting from mere numerical values to actionable insights. It is essential to acknowledge that while the tools and techniques discussed offer immense promise, they are not without challenges. The energy sector has its intricacies, and one-size-fits-all solutions may not always be ideal. Furthermore, the transition to data-driven decision-making requires infrastructural, cultural, and skill-based shifts, which might pose implementation challenges. Therefore, these study contributions provide a solid foundation and direction for future explorations.

7 Conclusion

The integration of neural networks and machine learning in the energy sector offers transformative opportunities for enhanced efficiency and decision-making. This chapter has emphasized the critical importance of adept data management, especially when dealing with energy-centric datasets, and the fine distinction between optimization strategies tailored for the sector. Data, being the backbone of machine learning endeavors, necessitates meticulous management techniques to ensure their integrity and relevance for energy applications. The choice of an optimizer, further, can significantly shape the performance and reliability of models catering to energy needs. Furthermore, hyperparameter tuning stands out as an indispensable element in achieving optimal model outcomes, especially in the dynamic context of energy challenges. As the energy sector continues its pursuit of innovation and sustainability, the insights and guidelines presented in this chapter aim to be a beacon for professionals navigating the intricate terrains of data and neural networks. Harnessing the potential of these techniques can undoubtedly pave the way for a more data-driven and efficient energy future.

References

1. Hagan, M.T., Demuth, H.B., Beale, M.H., Jesús, O.D.: Neural Network Design. Martin Hagan (2014)
2. Ramanathan, R., Ravindran, A.R., Mathirajan, M.: Multi-criteria decision making: an overview and a comparative discussion. In: Big Data Analytics Using Multiple Criteria Decision-Making Models. CRC Press (2017)
3. Danish, M.S.S., Senjyu, T., Sabory, N.R., Danish, S.M.S., Ludin, G.A., Noorzad, A.S., Yona, A.: Afghanistan's aspirations for energy independence: water resources and hydropower energy. Renew. Energy. **113**, 1276–1287 (2017). https://doi.org/10.1016/j.renene.2017.06.090
4. Danish, M.S.S., Senjyu, T.: Shaping the future of sustainable energy through AI-enabled circular economy policies. Circ. Econ. **2**(2), 100040 (2023)
5. Danish, M.S.S., Elsayed, M.E.L., Ahmadi, M., Senjyu, T., Karimy, H., Zaheb, H.: A strategic-integrated approach for sustainable energy deployment. Energy Rep. **6**, 40–44 (2020). https://doi.org/10.1016/j.egyr.2019.11.039
6. Danish, M.S.S., Senjyu, T., Ibrahimi, A.M., Ahmadi, M., Howlader, A.M.: A managed framework for energy-efficient building. J. Build. Eng. **21**, 120–128 (2019). https://doi.org/10.1016/j.jobe.2018.10.013
7. Ahmad, T., Ali, S., Basit, A.: Distributed renewable energy systems for resilient and sustainable development of remote and vulnerable communities. Philos. Trans. R. Soc. A Math. Phys. Eng. Sci. **380**, 20210143 (2022). https://doi.org/10.1098/rsta.2021.0143
8. Röder, M., Mohr, A., Liu, Y.: Sustainable bioenergy solutions to enable development in low- and middle-income countries beyond technology and energy access. Biomass Bioenergy. **143**, 105876 (2020). https://doi.org/10.1016/j.biombioe.2020.105876
9. Hofbauer, L., McDowall, W., Pye, S.: Challenges and opportunities for energy system modelling to foster multi-level governance of energy transitions. Renew. Sust. Energ. Rev. **161**, 112330 (2022). https://doi.org/10.1016/j.rser.2022.112330
10. Steg, L., Perlaviciute, G., Sovacool, B.K., Bonaiuto, M., Diekmann, A., Filippini, M., Hindriks, F., Bergstad, C.J., Matthies, E., Matti, S., Mulder, M., Nilsson, A., Pahl, S., Roggenkamp, M.,

Schuitema, G., Stern, P.C., Tavoni, M., Thøgersen, J., Woerdman, E.: A research agenda to better understand the human dimensions of energy transitions. Front. Psychol. **12**, 672776 (2021)
11. Cook, D., Davíðsdóttir, B., Gunnarsdóttir, I.: A conceptual exploration of how the pursuit of sustainable energy development is implicit in the genuine Progress indicator. Energies. **15**, 2129 (2022). https://doi.org/10.3390/en15062129
12. Husaini, D.H., Lean, H.H., Puah, C.-H., Affizzah, A.M.D.: Energy subsidy reform and energy sustainability in Malaysia. Econ. Anal. Policy. **77**, 913–927 (2023). https://doi.org/10.1016/j.eap.2022.12.013
13. Rodic-Wiersma, L.: Guidelines for national waste management strategies moving from challenges to opportunities. United Nations Environment Programme (UNEP), Geneve (2013)
14. Project Management Institute: A Guide to the Project Management Body of Knowledge (PMBOK Guide). Project Management Institute (2017)
15. Danish, M.S.S., Senjyu, T.: AI-enabled energy policy for a sustainable future. Sustainability. **15**(9), 7643 (2023)
16. Danish, M.S.S.: AI in energy: overcoming unforeseen obstacles. AI. **4**, 406–425 (2023). https://doi.org/10.3390/ai4020022
17. Drachman, D.A.: Do we have brain to spare? Neurology. **64**, 2004–2005 (2005). https://doi.org/10.1212/01.WNL.0000166914.38327.BB
18. McCulloch, W.S., Pitts, W.: A logical calculus of the ideas immanent in nervous activity. Bull. Math. Biol. **52**, 99–115 (1990). https://doi.org/10.1007/BF02459570
19. Omitaomu, O.A., Niu, H.: Artificial intelligence techniques in smart grid: a survey. Smart Cities. **4**, 548–568 (2021). https://doi.org/10.3390/smartcities4020029
20. Shehab, M., Abualigah, L., Omari, M., Shambour, M.K.Y., Alshinwan, M., Abuaddous, H.Y., Khasawneh, A.M.: Chapter 8: Artificial neural networks for engineering applications: a review. In: Elsheikh, A.H., Abd Elaziz, M.E. (eds.) Artificial Neural Networks for Renewable Energy Systems and Real-World Applications, pp. 189–206. Academic Press, Cambridge (2022). https://doi.org/10.1016/B978-0-12-820793-2.00003-3
21. Alanis, A.Y., Arana-Daniel, N., Lopez-Franco, C.: Artificial Neural Networks for Engineering Applications. Academic Press, St. Louis (2019)
22. Goodfellow, I., Bengio, Y., Courville, A.: Deep Learning. MIT Press, Cambridge (2016)
23. Hastie, T., Tibshirani, R., Friedman, J.: The Elements of Statistical Learning: Data Mining, Inference, and Prediction, 2nd edn. Springer, New York (2016)
24. De Wilde, P.: Neural Network Models. Springer, London (1997). https://doi.org/10.1007/978-1-84628-614-8
25. Mitchell, T.M.: Machine Learning. McGraw-Hill, New York (1997)
26. Loy, J.: Neural Network Projects with Python: the Ultimate Guide to Using Python to Explore the True Power of Neural Networks Through Six Projects. Packt Publishing, Birmingham (2019)
27. de Wilde, P.: Neural Network Models: Theory and Projects, 2nd edn. Springer, London; New York (1997)
28. Haykin, S.: Neural Networks and Learning Machines. Pearson, New York (2008)
29. Danish, M.S.S., Nazari, Z., Senjyu, T.: AI-coherent data-driven forecasting model for a combined cycle power plant. Energy Convers. Manag. **286**, 117063 (2023). https://doi.org/10.1016/j.enconman.2023.117063
30. Jia, D., Yang, L., Gao, X., Li, K.: Assessment of a new solar radiation nowcasting method based on FY-4A satellite imagery, the McClear model and SHapley additive exPlanations (SHAP). Remote Sens. **15**, 2245 (2023). https://doi.org/10.3390/rs15092245
31. Hissou, H., Benkirane, S., Guezzaz, A., Azrour, M., Beni-Hssane, A.: A novel machine learning approach for solar radiation estimation. Sustain. For. **15**, 10609 (2023). https://doi.org/10.3390/su151310609
32. Castillo-Rojas, W., Medina Quispe, F., Hernández, C.: Photovoltaic energy forecast using weather data through a hybrid model of recurrent and shallow neural networks. Energies. **16**, 5093 (2023). https://doi.org/10.3390/en16135093

33. Bentsen, L.Ø., Warakagoda, N.D., Stenbro, R., Engelstad, P.: Probabilistic Wind Park power prediction using Bayesian deep learning and generative adversarial networks. J. Phys. Conf. Ser. **2362**, 012005 (2022). https://doi.org/10.1088/1742-6596/2362/1/012005
34. Adli, H.K., Husin, K.A.K., Hanafiah, N.H.M., Remli, M.A., Ernawan, F., Wirawan, P.W.: Forecasting and analysis of solar power output from integrated solar energy and IoT system. In: 2021 5th International Conference on Informatics and Computational Sciences (ICICoS), pp. 222–226. IEEE, Semarang (2021). https://doi.org/10.1109/ICICoS53627.2021.9651831
35. Nielson, J., Bhaganagar, K., Meka, R., Alaeddini, A.: Using atmospheric inputs for artificial neural networks to improve wind turbine power prediction. Energy. **190**, 116273 (2020). https://doi.org/10.1016/j.energy.2019.116273
36. Chen, K.-S., Lin, K.-P., Yan, J.-X., Hsieh, W.-L.: Renewable power output forecasting using least-squares support vector regression and Google data. Sustain. For. **11**, 3009 (2019). https://doi.org/10.3390/su11113009
37. Qing, X., Niu, Y.: Hourly day-ahead solar irradiance prediction using weather forecasts by LSTM. Energy. **148**, 461–468 (2018). https://doi.org/10.1016/j.energy.2018.01.177
38. Sivaneasan, B., Yu, C.Y., Goh, K.P.: Solar forecasting using ANN with fuzzy logic pre-processing. Energy Procedia. **143**, 727–732 (2017). https://doi.org/10.1016/j.egypro.2017.12.753
39. Gutierrez-Corea, F.-V., Manso-Callejo, M.-A., Moreno-Regidor, M.-P., Manrique-Sancho, M.-T.: Forecasting short-term solar irradiance based on artificial neural networks and data from neighboring meteorological stations. Sol. Energy. **134**, 119–131 (2016). https://doi.org/10.1016/j.solener.2016.04.020
40. Abuella, M., Chowdhury, B.: Solar power forecasting using artificial neural networks. In: 2015 North American Power Symposium (NAPS), pp. 1–5. IEEE, Charlotte (2015). https://doi.org/10.1109/NAPS.2015.7335176
41. Dahmani, K., Dizene, R., Notton, G., Paoli, C., Voyant, C., Nivet, M.L.: Estimation of 5-min time-step data of tilted solar global irradiation using ANN (artificial neural network) model. Energy. **70**, 374–381 (2014). https://doi.org/10.1016/j.energy.2014.04.011
42. Marquez, R., Pedro, H.T.C., Coimbra, C.F.M.: Hybrid solar forecasting method uses satellite imaging and ground telemetry as inputs to ANNs. Sol. Energy. **92**, 176–188 (2013). https://doi.org/10.1016/j.solener.2013.02.023
43. Tabari, H., Kisi, O., Ezani, A., Hosseinzadeh Talaee, P.: SVM, ANFIS, regression and climate based models for reference evapotranspiration modeling using limited climatic data in a semi-arid highland environment. J. Hydrol. **444–445**, 78–89 (2012). https://doi.org/10.1016/j.jhydrol.2012.04.007
44. De Mauro, A., Greco, M., Grimaldi, M.: What is big data? A consensual definition and a review of key research topics. AIP Conf. Proc. **1644**, 97–104 (2015). https://doi.org/10.1063/1.4907823
45. Ball, G.H.: Data analysis in the social sciences: what about the details? In: Proceedings of the November 30--December 1, 1965, fall joint computer conference, part I, pp. 533–559. Association for Computing Machinery, New York (1965). https://doi.org/10.1145/1463891.1463950
46. Pöppelbaum, J., Chadha, G.S., Schwung, A.: Contrastive learning based self-supervised time-series analysis. Appl. Soft Comput. **117**, 108397 (2022). https://doi.org/10.1016/j.asoc.2021.108397
47. Dean, J., Ghemawat, S.: MapReduce: simplified data processing on large clusters. Commun. ACM. **51**, 107–113 (2008). https://doi.org/10.1145/1327452.1327492
48. Shrestha, Y.R., Krishna, V., von Krogh, G.: Augmenting organizational decision-making with deep learning algorithms: principles, promises, and challenges. J. Bus. Res. **123**, 588–603 (2021). https://doi.org/10.1016/j.jbusres.2020.09.068
49. Bera, S., Shrivastava, V.K.: Analysis of various optimizers on deep convolutional neural network model in the application of hyperspectral remote sensing image classification. Int. J. Remote Sens. **41**, 2664–2683 (2020). https://doi.org/10.1080/01431161.2019.1694725
50. Oh, S.-K., Kim, W.-D., Pedrycz, W., Joo, S.-C.: Design of K-means clustering-based polynomial radial basis function neural networks (pRBF NNs) realized with the aid of particle

swarm optimization and differential evolution. Neurocomputing. **78**, 121–132 (2012). https://doi.org/10.1016/j.neucom.2011.06.031
51. Vani, S., Rao, T.V.M.: An experimental approach towards the performance assessment of various optimizers on convolutional neural network. In: 2019 3rd International Conference on Trends in Electronics and Informatics (ICOEI), pp. 331–336 (2019), https://doi.org/10.1109/ICOEI.2019.8862686
52. Cochocki, A., Unbehauen, R.: Neural Networks for Optimization and Signal Processing. Wiley, New York (1993)
53. Mohapatra, R., Saha, S., Coello, C.A.C., Bhattacharya, A., Dhavala, S.S., Saha, S.: AdaSwarm: augmenting gradient-based optimizers in deep learning with swarm intelligence. IEEE Trans. Emerg. Top. Comput. Intell. **6**, 329–340 (2022). https://doi.org/10.1109/TETCI.2021.3083428
54. Gueorguieva, N., Valova, I., Klusek, D.: Solving large scale classification problems with stochastic based optimization. Procedia Comput. Sci. **168**, 26–33 (2020). https://doi.org/10.1016/j.procs.2020.02.247
55. Mirjalili, S., Mirjalili, S.M., Lewis, A.: Let a biogeography-based optimizer train your multi-layer perceptron. Inf. Sci. **269**, 188–209 (2014). https://doi.org/10.1016/j.ins.2014.01.038
56. Fulginei, F.R., Salvini, A., Parodi, M.: Learning optimization of neural networks used for MIMO applications based on multivariate functions decomposition. Inverse Probl. Sci. Eng. **20**, 29–39 (2012). https://doi.org/10.1080/17415977.2011.629047
57. Abdolrasol, M.G.M., Hussain, S.M.S., Ustun, T.S., Sarker, M.R., Hannan, M.A., Mohamed, R., Ali, J.A., Mekhilef, S., Milad, A.: Artificial neural networks based optimization techniques: a review. Electronics. **10**, 2689 (2021). https://doi.org/10.3390/electronics10212689
58. Danish, M.S.S.: A framework for modeling and optimization of data-driven energy systems using machine learning. IEEE Trans. Artif. Intell. **1–10** (2023). https://doi.org/10.1109/TAI.2023.3322395
59. Danish, M.S.S.: AI and expert insights for sustainable energy future. Energies. **16**, 3309 (2023). https://doi.org/10.3390/en16083309

Multidimensional Analysis and Optimization of Bus Loads for Enhanced Renewable Energy Integration in Power Systems

Mir Sayed Shah Danish, Soichiro Ueda, and Tomonobu Senjyu

1 Introduction

Analyzing the sensitivity of buses and identifying the critical ones in a power system offer numerous benefits [1]. It enhances system stability by pinpointing potential vulnerabilities to voltage or frequency fluctuations and aids in improved planning and design by highlighting buses with a significant impact on system performance [2]. This analysis is crucial for early fault detection, efficient resource allocation, and power flow optimization, reducing losses and improving efficiency. It plays a vital role in integrating renewable energy sources, contingency planning, and increasing grid flexibility. Furthermore, it assists in load balancing by ensuring no bus is overloaded and enhances the overall system reliability [3]. By focusing on critical buses, cost-effective system upgrades and maintenance can be achieved, and it also helps utilities meet regulatory requirements related to system reliability and performance [4].

Integrating renewable technology resources and storage systems can alter the load balance at various bus points in a power system. This change can directly impact system stability and reliability, mainly if it occurs at a sensitive bus [5]. Therefore, understanding the sensitivity of buses within a power system is crucial. It ensures system resilience and facilitates appropriate planning for future system expansion.

M. S. S. Danish (✉)
Energy Systems (Chubu Electric Power) Funded Research Division, Nagoya University, Nagoya, Japan
e-mail: danish@mdanish.me

S. Ueda · T. Senjyu
Department of Electrical and Electronic Engineering, University of the Ryukyus, Okinawa, Japan

M. S. S. Danish (ed.), *Unified Vision for a Sustainable Future*,
https://doi.org/10.1007/978-3-031-53574-1_2

Brucoli et al. [6] explored the sensitivity of stability limit curves against system parameter perturbations, aiming to define a more realistic and efficient margin for stability. The authors considered the impact of parameter uncertainty on the system's dynamic behavior and proposed a sensitivity analysis that relates the change of system eigenvalues to parameter changes. They also introduced a performance index variation as a measure of system response sensitivity to parameter changes.

Amusan et al. [7] presented a nodal analysis for identifying critical buses in the power grid and used a power simulation for load shedding. The authors developed a computational algorithm using differential evolution (DE) for minimizing service interruptions and blackouts and compared it against the conventional genetic algorithm (GA). The proposed algorithm was implemented on an IEEE 30-bus test system, and it showed an improvement in load-shedding efficiency after the application of DE.

Chureemart and Churueang [8] report the application of optimal power flow sensitivity for power system improvements. The authors proposed an optimal capacitor placement method that maximizes the system power factor, reduces total generation cost, and minimizes system losses. The authors applied sensitivity analysis to evaluate the change in the total generation cost with respect to the change in reactive power at any bus, determining the optimal location for capacitor banks. The proposed method was tested on 5-bus and 9-bus systems, demonstrating that the sensitivity technique yields the same results as more calculation-intensive methods.

Zhang et al. [1] proposed a load-shedding model based on sensitivity analysis to improve the efficiency and timeliness of online power system operation risk assessment. The authors proposed a method that calculates the sensitivity of each branch on each bus, identifying a collection of buses that significantly influence the reduction of power flow on overloaded branches. This transforms global optimization into local optimization, narrowing the solution range. The proposed model was tested on an IEEE 24-bus system and an IEEE 300-bus system.

Zhou and Zhu [9] introduced a sensitivity analysis framework to model the effects of bus split events on power grid security. They highlighted the potential for substation circuit breakers to be compromised, leading to bus split contingencies. Their framework uses the bus-branch representation to efficiently update system changes, treating a bus split as a line outage event followed by an additional power transfer. This model can be integrated into security-constrained economic dispatch problems to improve generation dispatch security. Their work offers an analytical approach to bus split modeling, thereby improving power system security in the face of potential bus split contingencies.

The reviewed studies present innovative approaches to improving power system stability, efficiency, and security through sensitivity analysis [10]. These studies collectively contribute to the advancement of power system analysis and operation, demonstrating the potential of sensitivity analysis in addressing complex power system challenges [11]. However, the present study advances the field of renewable energy accommodation by proposing a model based on time-sequential simulation, allowing for a more detailed and accurate representation of system dynamics [12]. A unique aspect of this study is the inclusion of a sensitivity analysis to identify critical factors (buses) affecting renewable energy accommodation, aiding in the prioritization of system improvements.

The chapter is organized as follows: Section 2 presents a case study, followed by a discussion on system sensitivity modeling in Sect. 3. Section 4 quantifies sensitivity and its impact on the integration of renewable energy in power systems. Section 5 introduces an online simulation tool. In Sect. 6, the optimization of the critical bus load is discussed to enhance system loadability and facilitate renewable integration. Section 7 delves into the data preparation and analysis processes. The architecture of the model is presented in Sect. 8, followed by the training strategy in Sect. 9. Section 10 discusses the formulation of the model, and Sect. 11 details the testing of the model. The results and discussion of the optimization process are presented in Sect. 12. Finally, Sect. 13 concludes the paper with a summary of the overall findings of the study.

2 Case Study

In this study, we focus on the Iowa 240 bus system, a fully observable radial distribution network located in the Midwest region of the United States, as our primary case study. The time-series Iowa 240 bus system (Fig. 1) is a radial distribution network comprising three feeders, all powered by a 69/13.8 kV substation. This actual grid system is located in the Midwest region of the United States and is the property of a municipal utility. It is fully observable and equipped with smart meters at all customer locations. The system features 240 primary network buses and a

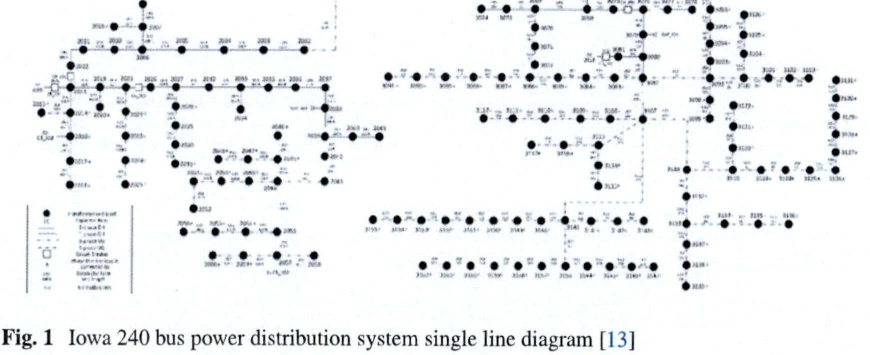

Fig. 1 Iowa 240 bus power distribution system single line diagram [13]

primary feeder conductor spanning 23 miles. This study investigates the first feeder with 15 loaded buses out of 17 buses as the case study.

3 System Sensitivity Modeling

Sensitivity analysis is the process of estimating how target variables vary in response to modifications in input variables [6, 14]. In the context of power systems, it examines the impacts of load changes on different load buses in terms of input variables (load changes) on the absolute quantities of target variables (the most sensitive buses to load changes) [15]. Sensitivity analysis in power systems addresses crucial questions regarding the system's reliability. It explores potential scenarios resulting from minor or major changes in the system's load patterns or distributed generations (DGs) [16]. It contributes to identifying the most appropriate strategies to enable planners and operators to handle uncertainties and prepare for expected or unexpected scenarios [17].

The Python script featured in Appendix 1 performs a nuanced exploration of a power system dataset [18]. This multifaceted analysis delves into sensitivity, correlation, regression, and Granger causality analyses [19, 20]. Each component of this study contributes to a broader understanding of interrelated phenomena within the system. Most notably, the script effectively identifies weak buses through an in-depth, all-encompassing investigation [7]. This rigorous analysis provides valuable insights through multidimensional analysis that differ from the literature.

The model script calculates the normalized standard deviation for each bus. This measure identifies buses that are most sensitive to load changes over time. Higher normalized standard deviations suggest a higher sensitivity to load changes [21]. By understanding this sensitivity, system operators can take preventive measures for highly sensitive buses and thus improve the system's overall reliability. In the context of a simple linear regression model, the sensitivity of the dependent variable y to the independent variable x can be represented by the slope of the regression line, $b1$, in Eq. 1 [22, 23]:

$$y = b0 + b1 * x + e \tag{1}$$

where $b0$ is the y-intercept, b1 is the slope of the line (representing the sensitivity of y to changes in x), and e is the error term.

The sensitivity analysis shown in Fig. 2, from approximately 42% to 95%, measures each bus's responsiveness to load changes. Bus B03 displays the utmost sensitivity (95%), indicating its significant susceptibility to even minor load fluctuations, followed by buses 13 and 11, with 89% and 86% sensitivity records, respectively. Bus B09, with its lowest sensitivity score of 42%, is the least influenced by load changes among the evaluated buses.

The correlation analysis examines the pairwise correlation between all numeric variables in the dataset, providing a comprehensive understanding of the

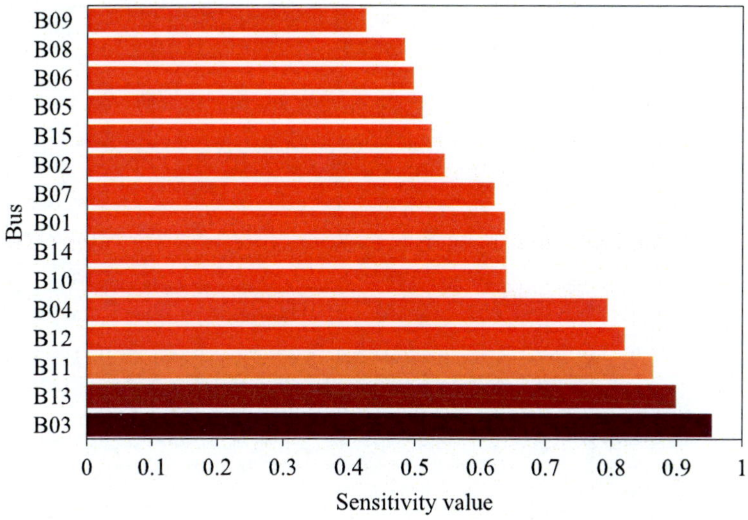

Fig. 2 Sensitivity values of various buses in the selected feeder of the power system

associations among different buses [24]. The correlation matrix shown in Fig. 3 visualizes these correlations, allowing a more straightforward interpretation. High correlations indicate buses where changes in load occur synchronously, which could have implications for power balancing and redundancy planning. The formula for Pearson's correlation coefficient r between the two variables x and y is given by Eq. 2 [12]:

$$r = \frac{\sum_{i-1}^{n}(x_i - \overline{x})(y_i - \overline{y})}{\sqrt{\sum_{i-1}^{n}(x_i - \overline{x})^2 \sum_{i-1}^{n}(y_i - \overline{y})^2}} \quad (2)$$

where n is the number of observations, x_i and y_i are individual observations of the variables x and y, respectively, and $\overline{x}$ and $\overline{y}$ are the means of x and y, respectively.

The correlation matrix presented in Fig. 3 visually depicts the relationships between load changes across various buses in the power system. Each value in the matrix, ranging from −1 to 1, represents the degree of correlation between two buses. A value of 1 indicates a perfect positive correlation, suggesting that as the load on one bus increases, so does the load on the other. A value of −1 implies a perfect negative correlation, meaning that an increase in the load on one bus corresponds to a decrease on the other. A value of 0 signifies no correlation. The diagonal of the matrix, filled with 1 s, highlights the perfect correlation of each bus with itself. Notably, there are strong positive correlations (between B01 and B03) and negative correlations (between B13 and B03), indicating coordinated and opposite load change behaviors, respectively. However, most matrix values are below 0.5, indicating generally weak correlations between load changes on different buses.

	B01	B02	B03	B04	B05	B06	B07	B08	B09	B10	B11	B12	B13	B14	B15
B01	1.00	0.80	0.83	0.78	0.11	0.82	0.21	0.07	0.28	0.14	0.39	0.19	0.02	0.14	0.19
B02	0.80	1.00	0.74	0.61	0.07	0.64	0.25	0.07	0.29	0.21	0.31	0.26	0.07	0.21	0.25
B03	0.83	0.74	1.00	0.76	0.04	0.69	0.10	0.02	0.19	0.07	0.28	0.13	-0.05	0.07	0.11
B04	0.78	0.61	0.76	1.00	0.07	0.66	0.03	0.01	0.14	0.05	0.28	0.08	-0.08	0.05	0.02
B05	0.11	0.07	0.04	0.07	1.00	0.18	0.22	0.50	0.40	0.31	0.27	0.22	0.34	0.31	0.21
B06	0.82	0.64	0.69	0.66	0.18	1.00	0.16	0.17	0.26	0.11	0.37	0.14	0.02	0.11	0.14
B07	0.21	0.25	0.10	0.03	0.22	0.16	1.00	0.23	0.40	0.39	0.20	0.25	0.33	0.39	0.33
B08	0.07	0.07	0.02	0.01	0.50	0.17	0.23	1.00	0.34	0.29	0.24	0.26	0.36	0.29	0.25
B09	0.28	0.29	0.19	0.14	0.40	0.26	0.40	0.34	1.00	0.50	0.26	0.35	0.40	0.50	0.39
B10	0.14	0.21	0.07	0.05	0.31	0.11	0.39	0.29	0.50	1.00	0.23	0.38	0.40	1.00	0.35
B11	0.39	0.31	0.28	0.28	0.27	0.37	0.20	0.24	0.26	0.23	1.00	0.20	0.19	0.23	0.21
B12	0.19	0.26	0.13	0.08	0.22	0.14	0.25	0.26	0.35	0.38	0.20	1.00	0.27	0.38	0.33
B13	0.02	0.07	-0.05	-0.08	0.34	0.02	0.33	0.36	0.40	0.40	0.19	0.27	1.00	0.40	0.36
B14	0.14	0.21	0.07	0.05	0.31	0.11	0.39	0.29	0.50	1.00	0.23	0.38	0.40	1.00	0.35
B15	0.19	0.25	0.11	0.02	0.21	0.14	0.33	0.25	0.39	0.35	0.21	0.33	0.36	0.35	1.00

Fig. 3 Correlation matrix of load changes across buses in the selected feeder in the power system

This reveals that in addition to the impact of connected buses on the most sensitive B03 bus, buses with no direct connection significantly impact the sensitivity of B03 with high impacts, such as B04 and B06.

The regression analysis is conducted to identify the relationships between the target variable and several other explanatory variables representing different bus loads. This regression model could predict one bus's load based on others' loads. The regression analysis results are visualized in Fig. 4, which compares the actual values of B03 with the predicted values from the regression model, providing a sense of the model's accuracy. Multiple linear regression is given by Eq. 3 [12]:

$$y = b0 + b1 * x1 + b2 * x2 + \ldots + bn * xn + e \tag{3}$$

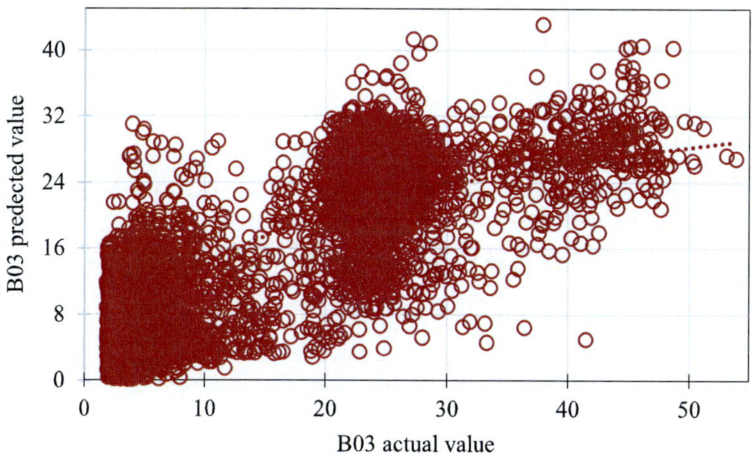

Fig. 4 The relationship between the actual target variable and the predicted values

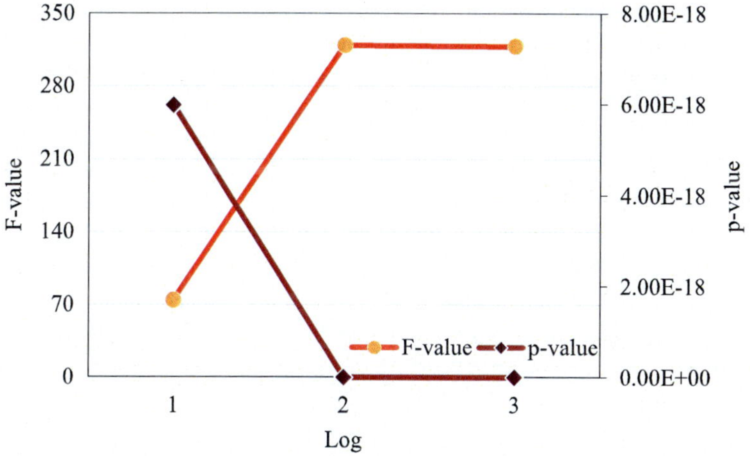

Fig. 5 Results of Granger causality analysis of the system

where *bo*, *b*1, ..., *bn* are the regression coefficients and *x*1, *x*2, ..., *xn* are the independent variables.

Finally, a Granger causality analysis hypothesis test determines whether one time series is useful in forecasting another. This code conducts a Granger causality test between B03 and B01 with up to 3 lags (Fig. 5). This analysis reveals if load changes in one bus can predict load changes in another, which could be vital for system planning and forecasting. In view of a two-variable system, x does not Granger-cause y if the coefficients of past values of x are all zero in a regression of y on past values of y and past values of x. This can be represented as follows (Eq. 4) [22, 23]:

$$y_t = a_0 + \sum_{i=1}^{n} a_i * y_{t-i} + \sum_{i=1}^{n} b_i * x_{t-i} + e_t \tag{4}$$

If all b_i are statistically equal to zero, then x does not Granger-cause y.

The Granger causality analysis results shown in Fig. 5 indicate a significant temporal relationship between the tested variables. At lag 1, an extremely low p-value (5.98E-18) and a high F-value (74.53) suggest that the independent variable significantly predicts the changes in the dependent variable. The significance intensifies at lags 2 and 3, as demonstrated by further diminished p-values (3.87E-70 and 8.64E-69, respectively) and an elevated F-value surpassing 300. These findings imply that the predictive power of the model improves markedly with the addition of more lags, reinforcing the conclusion that historical values of the independent variable are crucial in forecasting the future values of the dependent variable.

This model provides valuable insights into a power system's operation by analyzing bus load data from multiple angles. These analyses can provide information to power system operators, planners, and engineers about system sensitivities, relationships among bus loads, and cause-and-effect patterns. This information is a flexible resource to adopt for various analyses for effective renewable energy distribution planning, operation, and forecasting, ultimately enhancing renewable energy integration in power systems' efficiency and reliability.

The flowchart shown in Fig. 6 outlines an exhaustive data analysis process by quality assessment of the loaded data. If the data quality is poor, the function returns to the data loading stage or ends. If the data quality is good, the process proceeds with the functional process accordingly.

This study was conducted focusing on a single feeder to maintain simplicity. However, the proposed methodology is versatile and can be applied to the rest of the system or any power system scale [25]. Furthermore, it is adaptable to any scale of renewable resources integration, demonstrating its broad applicability in diverse scenarios.

4 Quantifying Sensitivity and Its Impact on Renewable Energy Integration in Power Systems

Sensitivity in a power system refers to how much a certain parameter (voltage or power) changes in response to a change in another parameter (load or generation). It is often quantified by calculating the partial derivatives of the power flow equations with respect to the parameters of interest.

The allocation of renewable energy resources in a power system can significantly affect the behavior of these sensitive buses. A large number of renewable energy resources is connected to a sensitive bus, which can lead to significant voltage fluctuations due to the intermittent nature of renewable energy resources. Therefore, power flow analysis can quantify the relationship between sensitive buses and

Fig. 6 Flowchart for the data analysis process including sensitivity, correlation, regression, and Granger causality analysis

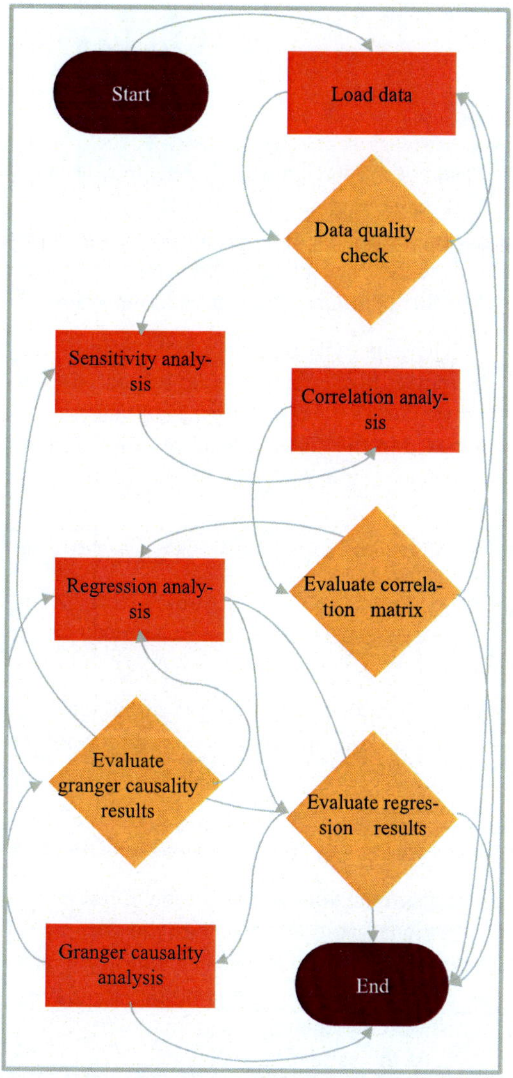

renewable energy allocation. The power flow equations govern the power distribution from generators to loads in a power system. These equations are typically nonlinear (Eqs. 5 and 6):

$$P_i = V_i \sum_{j=1}^{N} V_j \left(G_{ij} \cos\left(\theta_{ij}\right) + B_{ij} \sin\left(\theta_{ij}\right)\right) \tag{5}$$

$$Q_i = V_i \sum_{j=1}^{N} V_j \left(G_{ij} \sin\left(\theta_{ij}\right) - B_{ij} \cos\left(\theta_{ij}\right) \right) \tag{6}$$

where

P_i and Q_i are the real and reactive power at bus i.
V_i is the voltage magnitude at bus i.
G_{ij} and B_{ij} are the elements of the bus admittance matrix.
θ_{ij} is the voltage angle difference between buses i and j.
N is the total number of buses in the system.

The sensitivity of a bus can be quantified by calculating the partial derivatives of the power flow equations with respect to the bus power and voltage. These are often referred to as the power flow Jacobian matrix, a key component in power flow studies used in the Newton-Raphson method for solving the power flow problem. The Jacobian matrix J is given by Eq. 7:

$$J = \begin{bmatrix} \frac{\partial P}{\partial \delta} & \frac{\partial P}{\partial V} \\ \frac{\partial Q}{\partial \delta} & \frac{\partial Q}{\partial V} \end{bmatrix} \tag{7}$$

where P and Q are the vectors of the real and reactive power injections at each bus and δ and V are the vectors of voltage angles and magnitudes at each bus.

The elements of the Jacobian matrix represent the sensitivities of the power injections to changes in the power and voltage angles and magnitudes. For example, the element $\frac{\partial P_i}{\partial V_j}$ represents the sensitivity of the real power injection at bus i to a change in the voltage magnitude at bus j.

By analyzing these sensitivities, we can determine how changes in the power injections from renewable generation will affect the power and voltages at the sensitive buses. The developed online simulation tool is an effective way to evaluate sensitivity by changing and analyzing the most sensitive bus condition in the system. At the same time, such online simulation tools can also be developed for the rest of the sensitive buses. This can provide valuable information for planning and operating the power system, especially when integrating large amounts of renewable energy.

Optimization techniques can be used for a more detailed analysis to determine the optimal allocation of renewable resources in the system, which is not within the scope of this study. This can involve solving an optimization problem that minimizes the total generation cost or maximizes the total renewable generation, subject to the power flow equations and other system constraints.

5 Online Simulation Tool

The online simulation tool provided in Fig. 7 (available online: http://htmlpreview.github.io/?https://github.com/mirsayedshah/230704/blob/main/simulator_model.html) is a web-based simulator for predicting the future values of a system's most critical bus, which is denoted as "B03". The simulator takes into account the influence of the most impactful buses on B03 under two conditions: their value in a previous time period (t-1) and their steady-state value before the simulation.

The script is written in HTML, CSS, and JavaScript with the help of a neural designer. The script's HTML and CSS parts define the web page structure and style, respectively. The JavaScript part of the script is responsible for the functionality of

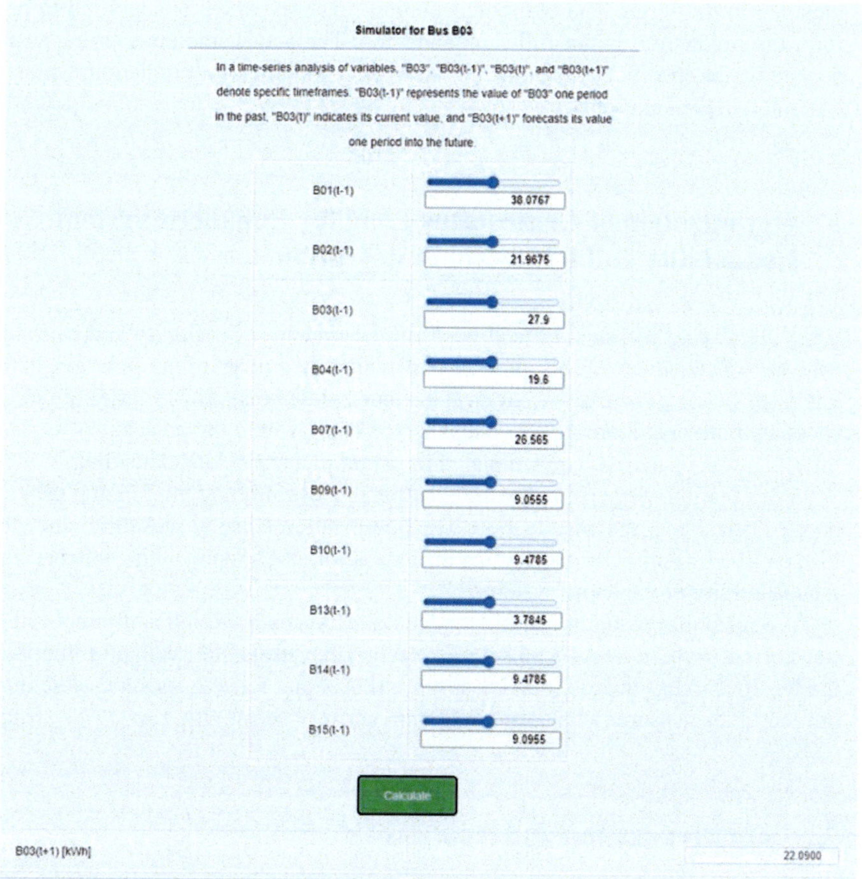

Fig. 7 Web-based simulator for predicting the future values of the critical B03 bus in a system (available online: http://htmlpreview.github.io/?https://github.com/mirsayedshah/230704/blob/main/simulator_model.html)

the simulator, which includes collecting input values, performing calculations, and displaying the results.

Its novelty lies in the application of time-series analysis in a web-based simulator. The simulator uses a neural network model to predict the future values of the system's most critical bus, taking into account the influence of other buses. This approach enables efficient resource allocation, facilitates the integration of renewable energy sources by managing variability, and maintains system stability by pinpointing potential vulnerabilities to voltage or frequency fluctuations.

This tool provides a user-friendly interface for performing complex time-series analysis compared to the existing literature. It allows users to adjust the values of various buses and instantly see the predicted future value of the critical bus. This interactive feature makes the script useful for research and educational purposes.

However, it is important to note that the actual neural network model is not included in the script. The output calculation function, which is supposed to perform the prediction, is not fully implemented. The actual implementation would depend on the specific complex neural network model used for prediction, which is described in the next section.

6 Optimization of Critical Bus Load for Enhanced System Loadability and Renewable Integration

After identifying the most critical buses in the system, a sensitivity and causality analysis was conducted. This analysis considered the impact of not only neighboring buses but also the entire system. It became evident that critical buses are influenced by both neighboring and nonconnected buses within the power system. This influence was visualized and simulated for a comprehensive understanding.

In this section, we perform an optimization process aimed at minimizing the load on the critical bus, specifically B03. The optimization strategy seeks to reduce the load on B03 by optimally distributing the load across the system, taking into account the loadability of the entire system.

As a result of this optimization, the load on B03 can be significantly reduced or transferred from an initial 11.12–10.83 kWh. This substantial reduction improves the bus loadability margin, making it more robust and reliable for renewable integration. It also ensures a balanced system in terms of sensitivity.

7 Data Preparation and Analysis

Time-series analysis is a statistical technique that deals with time-series data or trend analysis [26]. On the other hand, time-series forecasting involves the use of a model to predict future values based on previously observed values [27]. In this study, the time-series analysis of variables "B03," "B03(t-1)," "B03(t)," and "B03(t + 1)" denotes specific timeframes. "B03(t-1)" represents the value of "B03" one period in the past, "B03(t)" indicates its current value, and "B03(t + 1)" forecasts its value one period into the future.

7.1 Dataset Variables and Patterns

The dataset of the case study is structured into a 16-column, 8758-sample matrix required for constructing a predictive model. Columns signify variables, which can be classified into three types: independent (inputs), dependent (targets), and unused. This dataset has 31 inputs, 1 target, and 16 unused variables, all of which are numerically represented. Samples are divided into training (60%, or 5256 samples), selection (20%, or 1751 samples), and testing (20%, or 1751 samples) categories. Moreover, no unused samples are reported, and the dataset is entirely free of missing values. However, the number of filtered samples is 1, which accounts for less than 0.01% of the total samples in the dataset. The index of the unused sample is 8758. These samples are instrumental in model construction, selection, and validation.

7.2 Data Integrity and Statistical Parameters

In model construction, fundamental statistics are pivotal in identifying potential outliers within the dataset. An outlier is a data point in a dataset that significantly deviates from other observations, often due to variability in the data or measurement errors [28]. It is essential to meticulously verify key statistical parameters of each variable, encompassing minima, maxima, means, and standard deviations, as depicted in Fig. 8. These metrics offer insights into data integrity and robustness.

The minima refers to the smallest value within a dataset, while the maxima is the largest value. These two parameters provide a sense of the range of the data. The mean, or average, is calculated by summing all the values in the dataset and dividing by the count of values, providing a central value that can be representative of the dataset as a whole. The standard deviation (Eq. 8) measures the dispersion or variation within the dataset [12]. It indicates how far apart the values are from the mean. A low standard deviation suggests that the values are closely clustered around the mean, while a high standard deviation indicates a wider spread of values, implying greater variability in the data:

$$\text{Standard deviation} = \sqrt{\frac{\sum_{i=1}^{n}\left(x_i - \text{Mean}\right)^2}{n}} \tag{8}$$

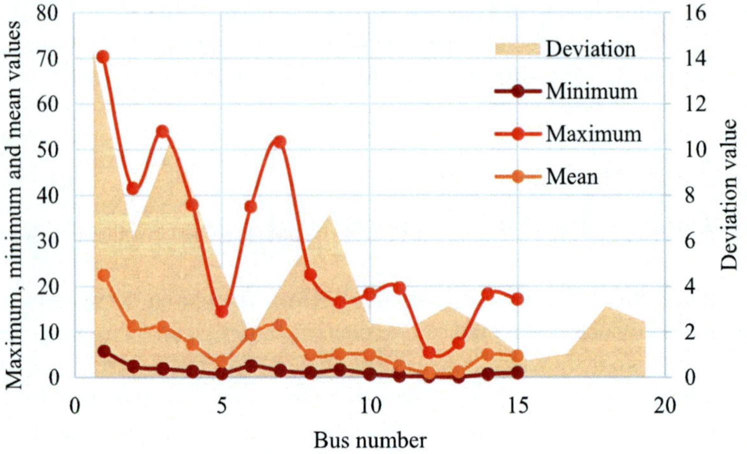

Fig. 8 Key statistical parameter analysis: minima, maxima, means, and standard deviations of variables

7.3 *Data Distributions*

Histograms shown in Fig. 9 display data distribution across their entire range and are essential in forecasting problems. A model's quality can be compromised by irregular data distributions. The accompanying charts present the histograms for the variables B01 to B15, with the horizontal axis representing bin centers and the vertical axis indicating the corresponding frequencies. The data manifest various frequency distributions. For instance, for variable B01, the highest frequency is 31.13% (2726 samples), centered at 9.004, and the lowest is 0.16% (14 samples) at 67.149. For B02, the maximum frequency is 41.49% (3634 samples), corresponding to 8.271, while the minimum is 0.1% (9 samples), located at 39.577. Similar patterns of varying frequencies are observed for the variables B03 to B15. This information showcases the unique distribution for each variable, reflecting the data diversity and highlighting the importance of considering these distributions in model development.

The ensuing charts elucidate box plots for the variables B01 to B15, outlining key measures such as minimum, first quartile (Q1), median (Q2), third quartile (Q3), and maximum. For B01, the values span from a minimum of 5.77 to a maximum of 70.38, with quartiles at 11.38 (Q1), 15.65 (Q2), and 33.51 (Q3). For the variable B02, the values range from 2.4 to 41.53, with respective quartiles at 6.69, 8.5, and 15.4. This pattern of distribution continues across all variables. For instance, B15 has a minimum of 1.01 and a maximum of 17.19, with quartiles at 2.75 (Q1), 4.01 (Q2), and 5.97 (Q3).

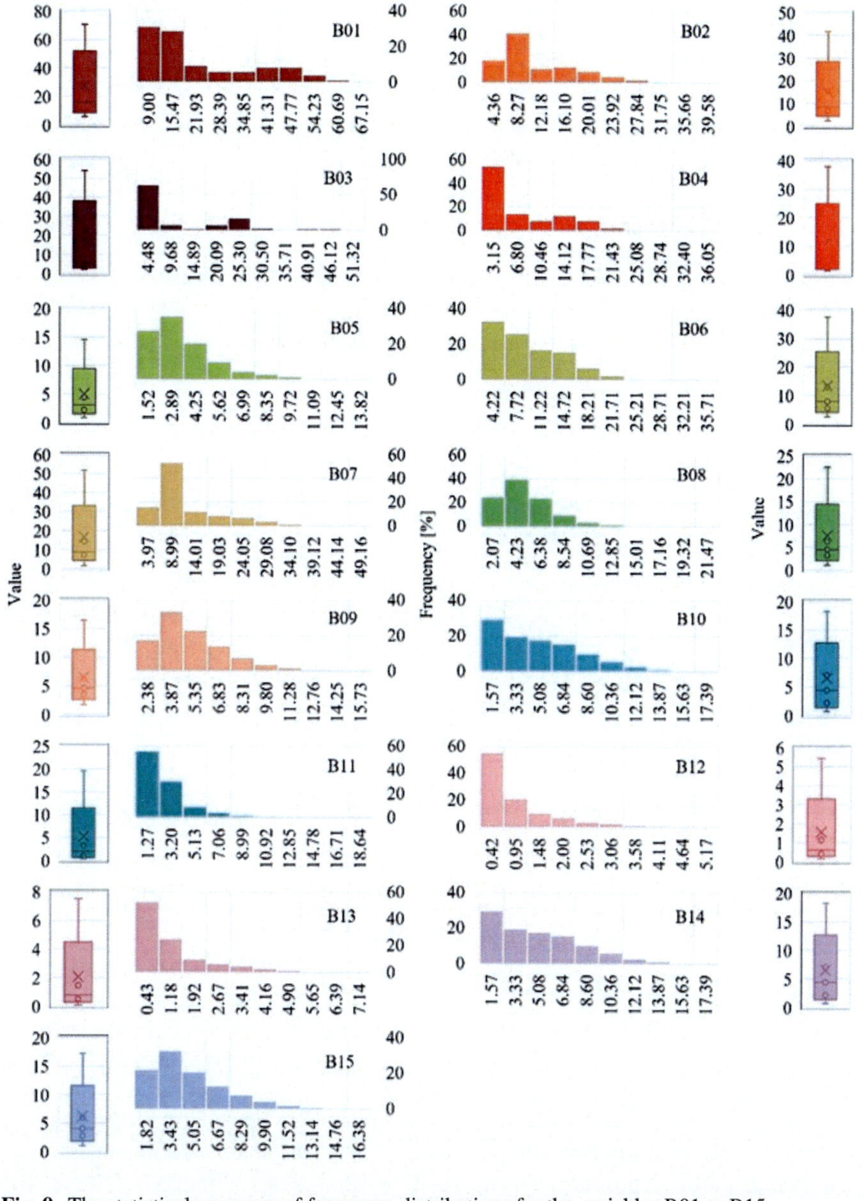

Fig. 9 The statistical measures of frequency distributions for the variables B01 to B15

7.4 Dependencies of Input and Output Variables

This section deals with the generation of scatter plots of the target variable (B03) against its corresponding input variables (the remaining 14 load buses) from a dataset to ascertain potential dependencies. Each plot involves a randomly selected

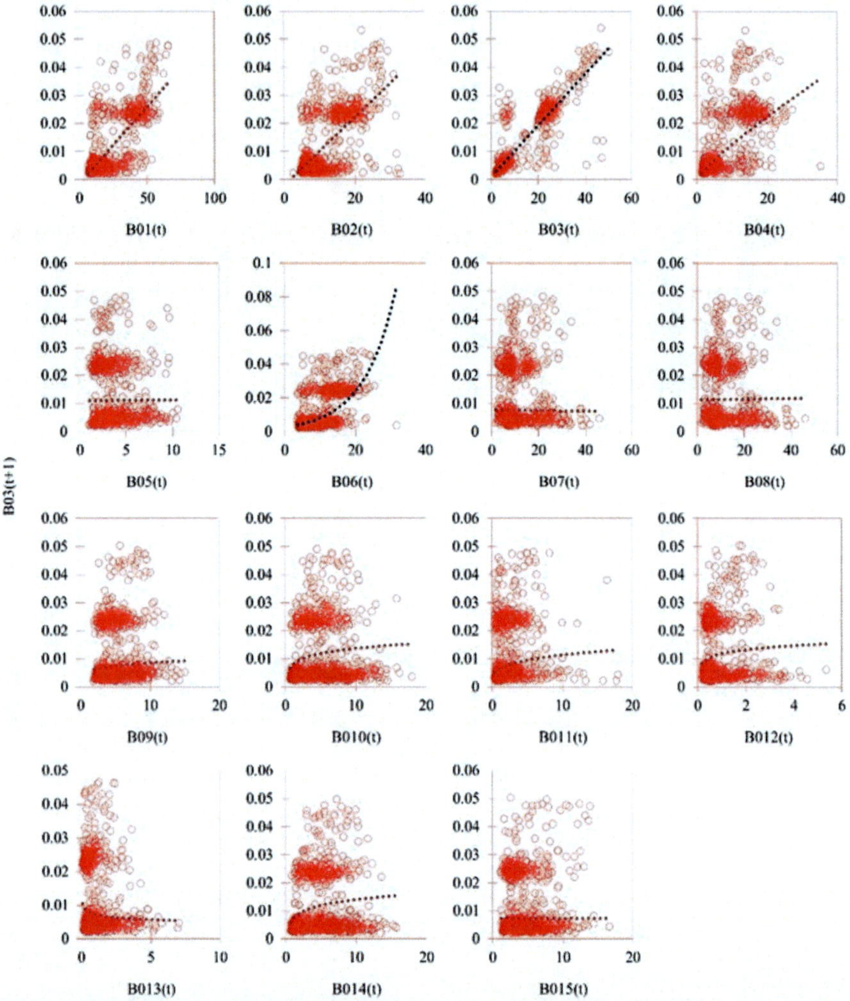

Fig. 10 Scatter plots of input variables against the target variable (B03) – relationships and correlations

subset of 1000 samples from the dataset and includes a regression line indicative of the type of relationship between the variables. The correlation coefficient associated with each plot also provides a measure of the strength and direction of the relationship [29].

The scatter plot of B01(t-1) vs. B03(t + 1) shown in Fig. 10 exhibits an exponential regression relationship with a higher correlation of 0.595. The B02(t-1) and B03(t + 1) graphs exhibit a linear relationship characterized by a correlation coefficient of 0.537. The B04(t-1) and B03(t + 1) plots also demonstrate a power relationship with a slightly lower correlation value of 0.629. In contrast, the B05(t-1) vs.

B03(t + 1) plot presents a linear relationship with a negative correlation of −0.032. The plots of B06(t-1), B07(t-1), and B15(t-1) against B03(t + 1) all exhibit exponential relationships with correlation values of 0.505, −0.079, and −0.079, respectively. The B08(t-1) vs. B03(t + 1) plot, on the other hand, presents a linear relationship with a negative correlation of −0.04. The scatter plots of B09(t-1), B11(t-1), and B13(t-1) against B03(t + 1) all exhibit power relationships with correlation values of 0.097, 0.222, and −0.145, respectively. Lastly, the B10(t-1), B12(t-1), and B14(t-1) vs. B03(t + 1) plots demonstrate logarithmic relationships with correlation values of 0.142, 0.078, and 0.142, respectively.

7.5 *Pearson's and Spearman's Correlations*

Pearson's correlation coefficient (r) is a statistical measure that calculates the linear relationship between two data points given in Eq. 2. It is computed using the individual data points, their means, and the total number of data points.

On the other hand, Spearman's correlation is a nonparametric method used to measure the monotonic relationship between two variables that is not restricted to linear relationships (Eq. 9) [12]. The Spearman correlation coefficient (ρ) is calculated using the difference between the ranks of observations (d_i) and the total number of data points (n) as follows:

$$\rho = 1 - \frac{6\sum d_i^2}{n\left(n^2 - 1\right)} \tag{9}$$

This section visualizes the computation of Pearson's and Spearman's correlation values among all input variables. This process is crucial for identifying the strength and nature of interrelationships within the dataset, which can assume various functional forms such as linear, exponential, power, logarithmic, or logistic. Correlation values, bound between 0 and 1, indicate the strength of relationships: a value close to 1 implies a strong relationship, while a value close to 0 signifies a negligible one.

Table 1 outlines the 10 most significant Pearson's correlations among variables. This figure highlights the model's three most substantial correlations: between B10 (t-1) and B14 (t-1), B10 (t-0) and B14 (t-0), and B01 (t-1) and B01 (t-0) with the corresponding values of 1, 1, and 0.932. This table also shows the 10 most substantial Spearman correlations among the inputs. This figure identifies the top three correlations between B10 (t-1) and B14 (t-1), B10 (t-0) and B14 (t-0), and B06 (t-1) and B06 (t-0) with the correlation coefficients of 1, 1, and 0.924, respectively.

Figure 11a provides a detailed enumeration of all correlations among the input variables, furnishing a holistic view of the variable interactions in the dataset. Fig. 11b depicts the Spearman correlations among all input variables, providing an alternative correlation metric for comprehensive analysis.

Table 1 The top 10 significant (a) Pearson's and (b) Spearman's correlations among the top 10 variables

No.	Input variable		Correlation value	
			Pearson	Spearman
1	B10(*t*-1)	B14(*t*-1)	1	1
2	B10(*t*)	B14(*t*)	1	1
3	B01(*t*-1)	B01(*t*)	0.932	0.923
4	B07(*t*-1)	B07(*t*)	0.927	0.919
5	B06(*t*-1)	B06(*t*)	0.925	0.924
6	B03(*t*-1)	B03(*t*)	0.908	0.898
7	B02(*t*-1)	B02(*t*)	0.903	0.896
8	B04(*t*-1)	B04(*t*)	0.892	0.876
9	B03(*t*-1)	B01(*t*)	0.879	–
10	B01(*t*-1)	B03(*t*-1)	0.831	–
11	B11(*t*-1)	B11(*t*)	–	0.825
12	B05(*t*-1)	B05(*t*)	–	0.812

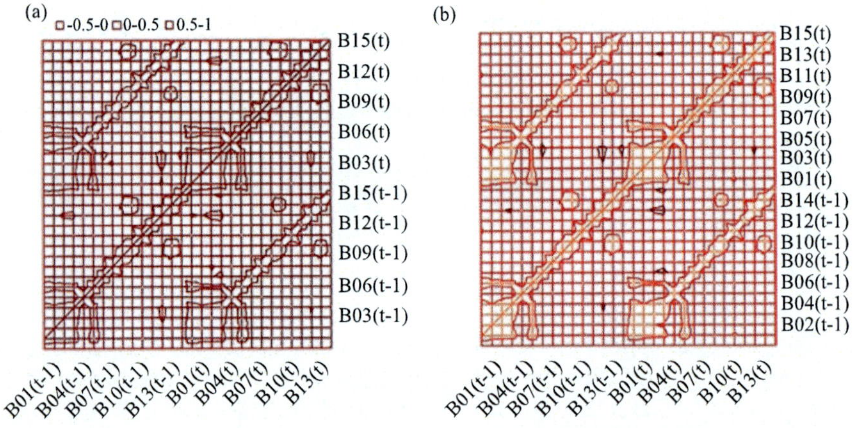

Fig. 11 Comprehensive overview of the (**a**) Pearson's and (**b**) Spearman's correlations among the input variables

Figure 12 presents Pearson's and Spearman's correlation coefficients for different variables at different time steps (*t*-1, *t*). Pearson's correlation measures the linear relationship between two datasets, and the Spearman correlation assesses how well the relationship between two variables can be described using a monotonic function.

Potential dependencies between the individual inputs and the targets within the dataset are investigated. Correlation coefficients are calculated for all inputs and targets, with coefficients close to −1 or 1 signifying a strong relationship and those close to 0 indicating negligible relationships, while targets typically depend on multiple inputs simultaneously (Fig. 13).

The cross-correlation shown in Fig. 14 presents serial correlations between variables at varying lags (*t*-1), commonly utilized in forecasting to explore a time-series

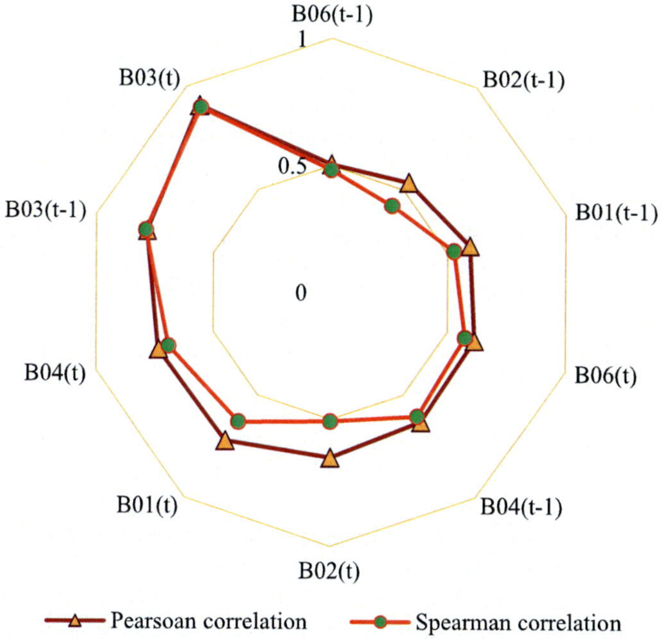

Fig. 12 Comparison of Pearson's and Spearman's correlation coefficients for different variables at different time steps

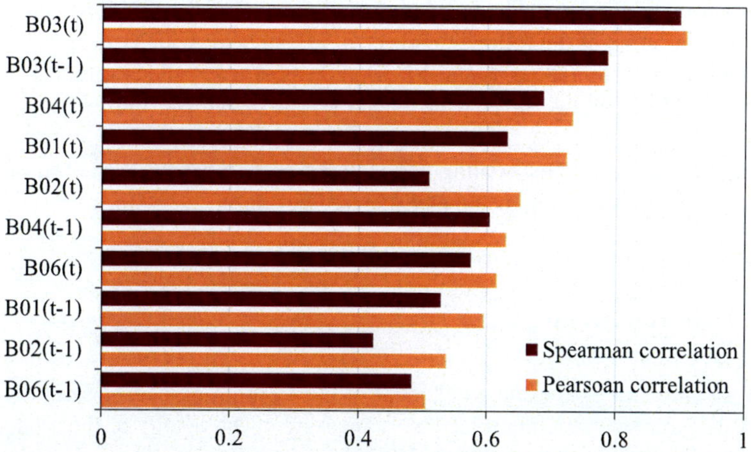

Fig. 13 Pearson's and Spearman's dependence of B03(t + 1) on the top 10 highly correlated input columns

(t) correlation with its past (t-1) and future values (t + 1). A positively correlated series exhibits persistence, where positive and negative deviations tend to be followed by similar deviations. Conversely, negative correlations typically see positive deviations succeeded by negative ones, and vice versa. In a random series, these

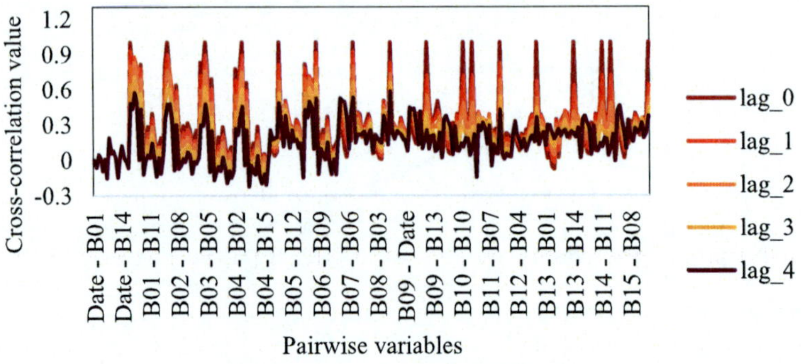

Fig. 14 Analysis of cross-correlation and serial correlations in time-series data for forecasting

correlations should hover near zero on average. For example, looking at the row "Date-B01," the cross-correlation at lag 0 is −0.0619131, which indicates a slight negative correlation between "Date" and "B01" variables when there is no shift. As the lag increases, the correlation coefficient changes, indicating how the correlation evolves as one series is shifted relative to the other.

8 Model Architecture

The utilized predictive model is a deep neural network, a class of universal approximators consisting of the following layers: a scaling layer with 31 neurons, two perceptron layers with 3 neurons and 1 neuron, respectively; an unscaling layer with 1 neuron; and a bounding layer with 1 neuron. The network is fed by 31 inputs and produces 1 output.

Figure 15 presents the scaling parameters for the 31 input variables, including the minimum, maximum, mean, and standard deviation.

9 Training Strategy

The training strategy, a crucial element of the learning process, refines the neural network parameters to minimize loss, a measure of the quality of the model's predictions. The loss index is contingent on two factors: error and regularization. The normalized squared error (NSE), chosen as the error measure, quantifies the model's data fit, where a value of 1 corresponds to a mean prediction and 0 to a perfect prediction. Regularization, implemented via L2 regularization, controls parameter values to promote model smoothness and prevent overfitting.

This strategy employs a quasi-Newton optimization algorithm, circumventing second-derivative calculations by approximating the inverse Hessian using gradient

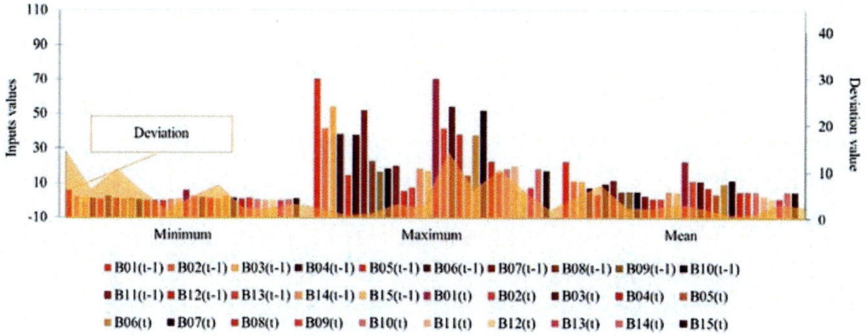

Fig. 15 Scaling parameters for 31 input variables: minimum, maximum, mean, and standard deviation

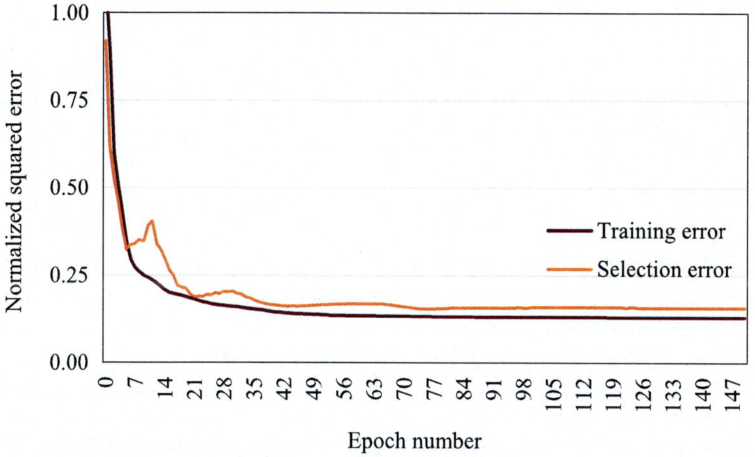

Fig. 16 Training and selection process error history

information. The method parameters are: inverse Hessian approximation method (BFGS), learning rate method (BrentMethod), learning rate tolerance (0.001), minimum loss decrease (0), loss goal (0.001), maximum selection error increases (100), maximum epoch number (1000), and maximum time (1).

The process of training the neural network using the quasi-Newton method focuses on adjusting the parameters of the network to minimize loss, using gradient information to approximate the inverse Hessian at each iteration, thereby obviating the need for second-derivative computations. Over the course of 150 epochs (Fig. 16), the initial training and selection errors of 1.1049 and 0.9216, respectively, reduce to 0.1313 and 0.1604, indicating effective model training. Key training outcomes are listed in Table 2, which includes the final states of the neural network, the loss index, and the optimization algorithm. The ultimate objective of the training strategy is to achieve a low selection error, a critical measure of the model's ability to generalize to new unseen data rather than just memorizing the training data.

Table 2 Training process results considering the loss index and the optimization algorithm

Parameter	Value
Epoch number	150
Elapsed time	00:00:04
Stopping criterion	Minimum loss decrease
Training error	0.131
Selection error	0.16

10 Model Formulation

Model selection algorithms aim to identify a neural network topology that minimizes error on novel data. These algorithms can be categorized into neurons selection algorithms, which determine the optimal number of hidden neurons, and input selection algorithms, which identify the optimal subset of input variables.

In this study, the growing neurons algorithm was utilized to ascertain the optimal number of neurons. Beginning with a minimum number of neurons, this method adds a defined quantity at each iteration. Table. 3 lists the parameters of the growing neurons algorithm used in this scenario, such as the minimum and maximum neurons, the number of trials for each neural network, and the maximum time allotted for the algorithm.

The outcome of this selection process is shown in Fig. 17 in which the training and selection errors across different subsets during the growing neurons selection process are graphically represented. The lowest selection error was observed with a model of nine neurons.

The network architecture after optimization contains scaling, perceptron, unscaling, and bounding layers with 31, 9, 1, and 1 neurons, respectively. The revised neural network architecture boasts an optimal neuron count as prescribed by the selection algorithm, consisting of 31 inputs, 1 output, and 9 neurons.

The growing inputs algorithm employed in this study identifies the optimal set of inputs for the application by sequentially incorporating the inputs based on their correlation with the target. This method is particularly useful in eliminating redundant inputs, which can impair the performance of the neural network, by finding the optimal subset of inputs that minimizes the model error. The algorithm given in Table 4 is designed to optimize the selection of inputs by running a specified number of trials for each network, aiming for a certain selection error goal, and limiting the number of iterations where the selection error increases. The algorithm also sets boundaries on the number of inputs in the network and the range of correlations to be considered. The process is constrained by a maximum number of iterations and a time limit.

Figure 18 presents the top 10 instances of maximal errors in the B03(t + 1) predictions. For each instance, it provides the index, the value of the maximal error, the sample in which this error occurred, and the corresponding values for 15 different datasets in that sample.

Table 3 Parameters and values for the neuron selection algorithm in neural network optimization

Parameter	Description	Value
Minimum neurons	Minimum number of hidden perceptrons to be evaluated	1
Maximum neurons	Maximum number of hidden perceptrons to be evaluated	10
Step	Number of hidden perceptrons added in each iteration	1
Trial number	Number of trials for each neural network	3
Selection loss goal	Goal value for the selection error	0
Maximum selection failures	Maximum number of iterations at which the selection error increases	100
Maximum iteration number	Maximum number of iterations to perform the algorithm	1000
Maximum time	Maximum time for the neurons selection algorithm	3600

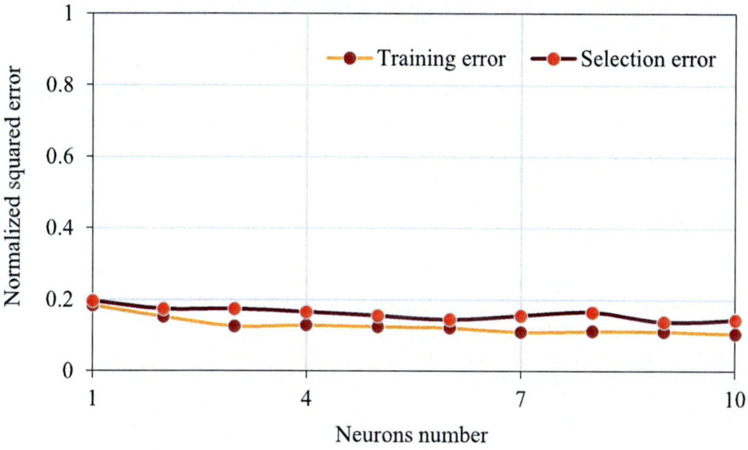

Fig. 17 Training and selection errors across different subsets during the growing neurons selection process

Table 5 presents the results of the input selection performed by the growing inputs algorithm, including the final states of the neural network, the loss index, and other associated parameters. The optimal subset of inputs ascertained by the algorithm is enumerated in the subsequent list.

The resultant neural network architecture now incorporates the optimal number of inputs suggested by the input selection algorithm. Composed of scaling, perceptron, unscaling, and bounding layers with 17, 9, 1, and 1 neurons, respectively, this optimized neural network structure accommodates 17 inputs, 1 output, and 9 neurons.

Table 4 Parameters and constraints for the neural network input selection algorithm

Parameter	Description	Value
Trial number	Number of trials for each neural network	3
Selection error goal	Goal value for the selection error	0
Maximum selection failures	Maximum number of iterations at which the selection error increases	100
Maximum input number	Maximum number of inputs in the neural network	31
Minimum correlations	Minimum value for the correlations to be considered	0
Maximum correlations	Maximum value for the correlations to be considered	1
Maximum iteration number	Maximum number of iterations to perform the algorithm	1000
Maximum time	Maximum time for the input selection algorithm	3600

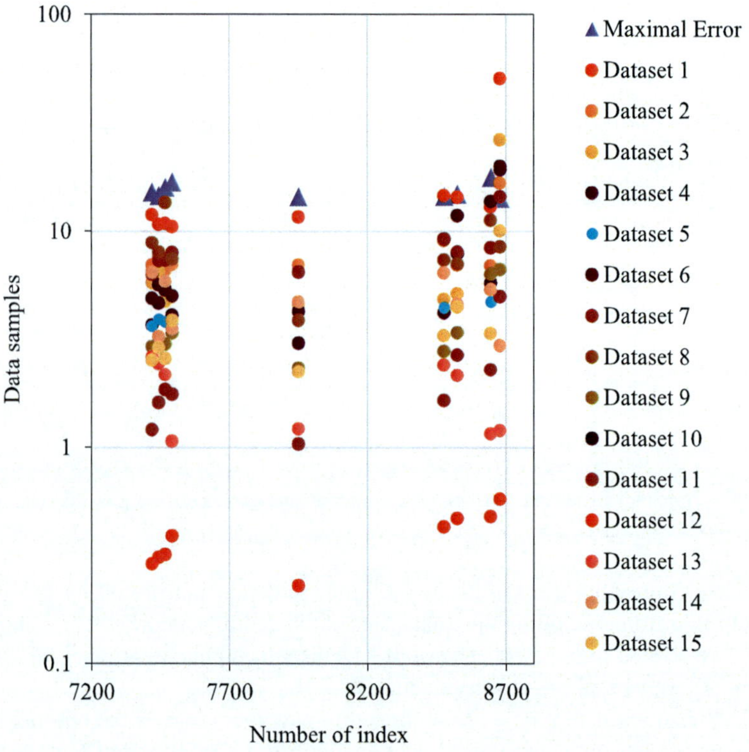

Fig. 18 Top 10 instances of maximal errors in B03(t + 1) predictions across multiple datasets

Table 5 Results and performance metrics of the growing neurons algorithm in neural network optimization

Parameter	Value
Optimal order	9
Optimum training error	0.1115
Optimum selection error	0.1383
Epoch number	10
Stopping criterion	Maximum neurons
Elapsed time	00:01:05

11 Model Testing

The goodness-of-fit is a statistical measure that quantifies how accurately a model's predictions align with the actual observations. Among the various metrics, the coefficient of determination, or R2, is widely employed. This coefficient gauges the proportion of the variation in the predicted variable that can be explained by the model, with a value of 1 signifying a perfect fit, that is, the predicted outputs perfectly match the target values. Figure 19 provides a comparative analysis of the predicted versus actual values for the output B03(t + 1). The optimal prediction scenario, where the outputs equate to the targets, is represented by the line.

This process quantifies the model's errors across all utilized samples, individually evaluating the model's performance at each instance. Figure 20 presents a comprehensive summary of these errors for each sample application.

11.1 Error Statistics

Error statistics, encompassing minima, maxima, means, and standard deviations of discrepancies between the neural network predictions and the testing data, serve as critical indicators of model quality. In the testing dataset, the percentage errors range from 0.000952831% to 34.1495%, with an average of 3.8548% (Table 6). Figure 21 outlines these statistics in more detail, comparing the relative error of a baseline model (using the initial values as predictions) with that of the neural network model.

11.2 Error Histograms

The error histograms provide insight into the distribution of prediction errors made by the neural network on the testing data. Ideally, each output variable would exhibit a normal error distribution. Figure 22 illustrates this distribution for the output B03(t + 1). The x-axis denotes the center of bins and the y-axis represents the corresponding frequencies. Notably, the bin centered at 0% has the highest frequency of 42.41%, while the bin with the lowest frequency (0.2%) is centered at −26.973%.

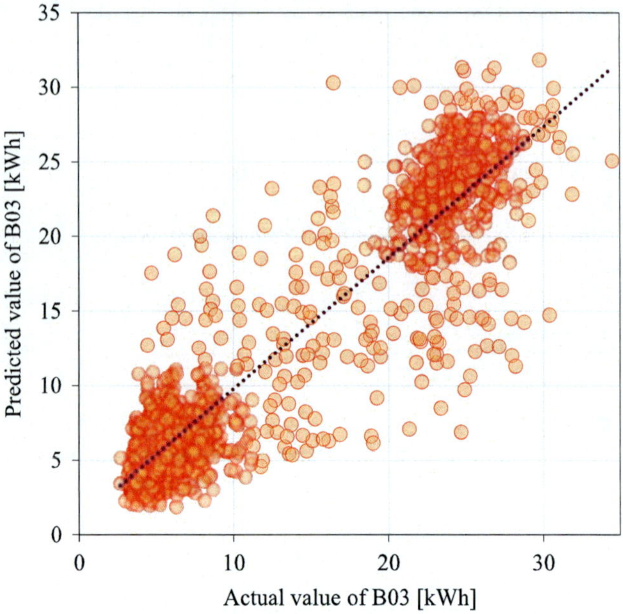

Fig. 19 Goodness-of-fit for B03 predictions of actual vs. predicted values

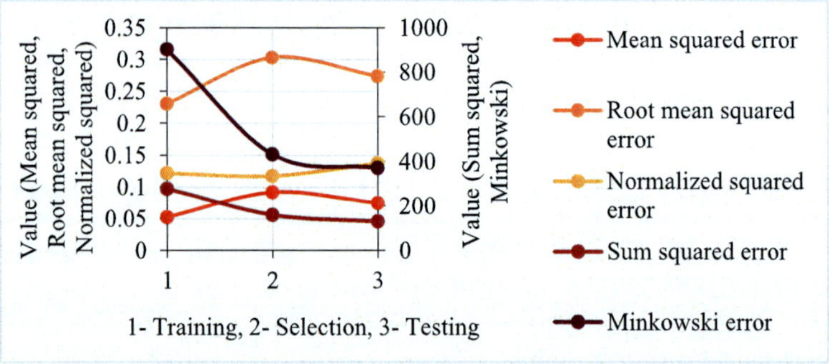

Fig. 20 Model performance evaluation of error metrics for training, selection, and testing datasets

Table 6 Comparison of error rates of a baseline model vs neural network for B03(t + 1)

Parameter	Relative error	Error (%)
Baseline model	0.0569982	5.69982
Neural network	0.038548	3.8548

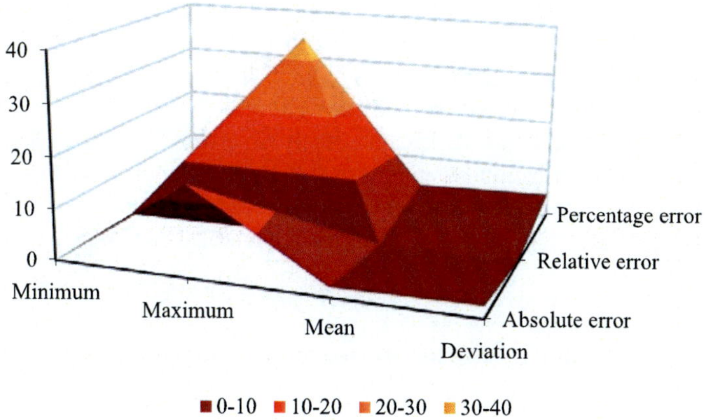

Fig. 21 B03(t + 1) predictions of error statistics summary

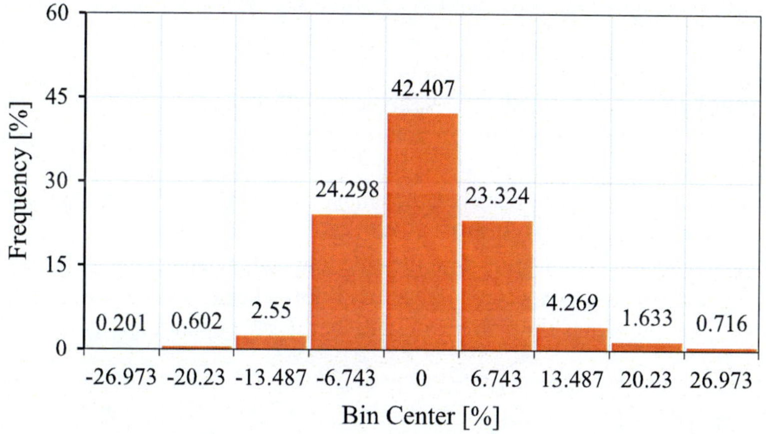

Fig. 22 Distribution of prediction errors for output B03(t + 1) by neural network on testing data, highlighting high accuracy and minimal underestimation

This suggests that the majority of predictions are closely aligned with the actual values, with fewer instances of substantial underestimation.

Identifying the test samples that yield the maximum errors can be extremely useful. It highlights potential weaknesses in the predictive model, suggesting areas where the model's predictions deviate significantly from the actual values. This analysis can guide further refinement of the model, thereby improving its overall predictive accuracy and robustness.

11.3 Time-Series Plot

This analysis provides a graphical depiction of the time-series data from both the target dataset and the model's predictions (Fig. 23). By plotting observations (y-axis) against time (x-axis), it can visually assess the model's accuracy. Such visual representation facilitates an intuitive understanding of how closely the model's predictions align with the actual target data over time. It can reveal patterns or discrepancies that might be less obvious from numerical error metrics alone.

11.4 Importance of Inputs at the System Level

Identifying the variables that most influence a given prediction is critical. This can be accomplished by computing the derivatives of a model's outputs with respect to its inputs (Fig. 24). A high derivative value suggests a significant impact from a variable, while a value close to zero indicates less influence.

Figure 25 ranks the input variables impacting the output variable B03(t + 1) based on their values. A positive value indicates that an increase in the input variable will correspondingly increase the output variable. In contrast, a negative value suggests that an increase in the input variable will decrease the output variable. Near-zero values imply a minimal change in the output variable despite changes in the input variable.

According to the data presented, the model's most significant influences are from the three variables B03(t), B03(t-1), and B01(t-1). B03(t) shows a value of 0.328, suggesting a direct correlation with the outcome variable, that is, as it increases, the output variable is also expected to increase. B03(t-1), with a value of −0.199, exhibits an inverse relationship with the output, that is, as the input increases, the output

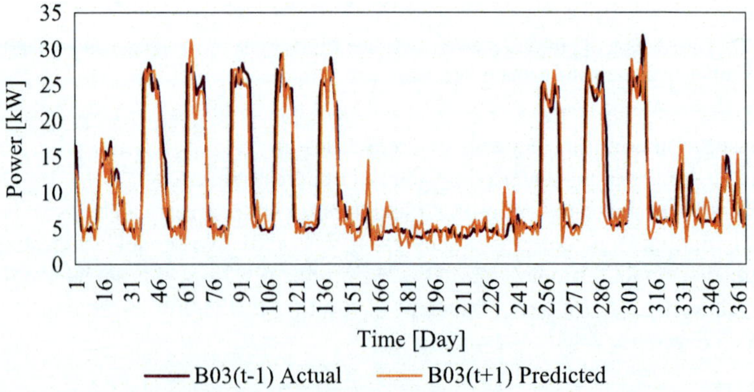

Fig. 23 Time-series plot comparing model predictions with the actual dataset over time

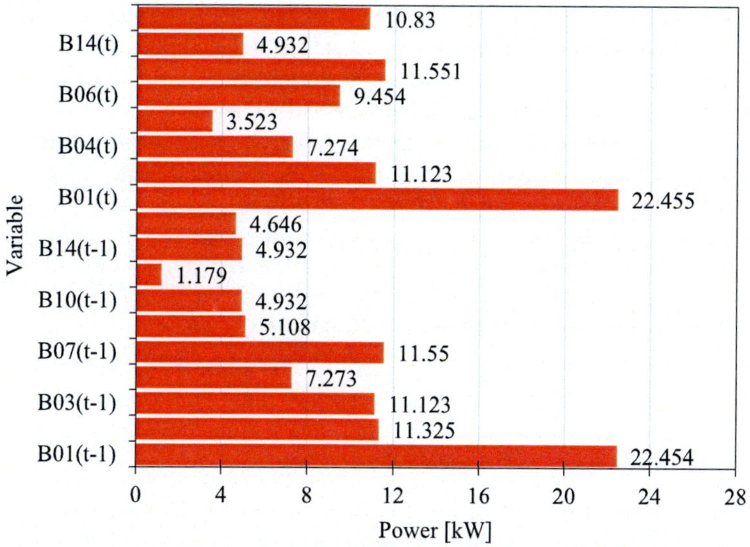

Fig. 24 Input–output analysis for the system variables

tends to decrease. Lastly, B01(t-1) carries a value of 0.103, which implies a

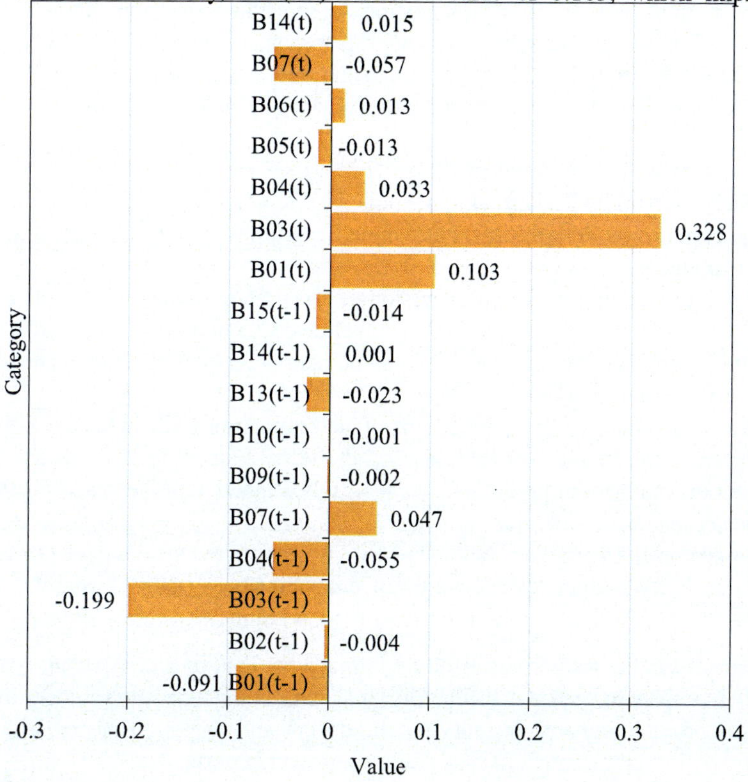

Fig. 25 Importance of input variables for predicting the output

moderate positive influence on the output variable, that is, an increase in B01(*t*-1) would result in a somewhat proportional increase in the output.

The system's output is significantly influenced by B03(*t*), B03(*t*-1), and B01(*t*-1). B03(*t*) positively affects the output, while B03(*t*-1) negatively impacts it. B01(*t*-1) also contributes positively, but its impact is relatively moderate. This reaffirms that B03 is the most sensitive in terms of output variation within the system's feeder network.

12 Optimization Discussion and Summary

This study aimed to optimize the load on a critical bus (B03) in the power system to enhance system loadability and facilitate renewable integration. Here are the key results and findings:

- Data analysis: A dataset structured into a 16-column, 8758-sample matrix was used for a time-series analysis. The number of filtered samples was less than 0.01% of the total samples, indicating high data integrity.
- Variable interrelationships: Scatter plots revealed various types of relationships between the variables, including exponential, linear, power, and logarithmic. The strength of these relationships was measured using correlation coefficients, with the 10 most significant Pearson's and Spearman's correlations ranging from perfect (1) to strong (0.924 for Spearman and 0.932 for Pearson).
- Predictive model: A deep neural network with 31 inputs and 1 output was used. The training strategy employed a quasi-Newton optimization algorithm, effectively reducing the initial training and selection errors.
- Model selection: The growing neurons algorithm identified an optimal model of nine neurons, and the growing inputs algorithm identified the optimal set of inputs.
- Error rate comparison: A comparison of error rates between the baseline model and the neural network for B03($t + 1$) showed a relative error of 5.69982% for the baseline model and 3.8548% for the neural network. This indicates that the neural network model had a lower error rate than the baseline model.
- Model evaluation: The coefficient of determination (R2) indicated a good fit between the model's predictions and the target values. Error statistics showed that the majority of predictions were closely aligned with the actual values, with fewer instances of substantial underestimation.
- Load optimization: The load on bus B03 was predicted to be reduced from 11.12 to 10.83 kWh (approximately 2.61%), enhancing the system's loadability margin and making it more robust for renewable integration.

Therefore, the study successfully optimized the load on a critical bus in a power system, enhancing system loadability and facilitating renewable integration. The results provide valuable insights for the design and operation of power systems, particularly in the context of renewable energy integration.

13 Conclusion

This study developed and implemented an innovative framework that extends beyond the traditional focus of identifying weak or sensitive buses within a power system. This approach provides a more comprehensive understanding of system load changes and resilience by examining the interrelationships and causal impacts among buses, including those not directly connected. The multidimensional nature of this methodology allows system planners and operators to visualize and anticipate the ripple effects of a load change on a single bus across the entire system. This is a significant advancement over previous models that only considered the impact on neighboring buses.

Furthermore, the study demonstrated the practical application of this approach in optimizing the load on the critical bus B03, reducing its load from 11.12 to 10.83 kWh. This optimization not only improved the loadability margin of the bus but also enhanced the robustness and reliability of the system for renewable integration.

The adaptability and effectiveness of this approach have been confirmed through experimental analysis of a sizable dataset from Iowa's 240-bus power system, suggesting its potential for wide-ranging applications in diverse power system scenarios. This study presents a significant step forward in power system analysis and planning, offering a robust and flexible tool that can help facilitate the transition toward more sustainable and resilient power systems.

Appendix 1

```
"""
Copyright © 2023, Danish M.
This work is licensed under a Creative Commons Attribution 4.0
International License. This license allows others to distribute, remix,
adapt, and build upon this work, even commercially, as long as they
credit the original creation.
For more information about this license, please visit https://
 corresponding to the first feeder's 15 load bus data from the Iowa 240
bus test system. creativecommons.org/licenses/by/4.0/
For any use beyond the scope of this license, please contact the
copyright holder at danish@mdanish.me.
"""

# This script is designed to be executed within the Jupyter Lab
environment, a web-based interactive development interface used for
Python programming.
import pandas as pd
import seaborn as sns
import matplotlib.pyplot as plt
import statsmodels.api as sm
import warnings

# The script loads the dataset
# The feeder 1 dataset can be found from this url:
https://github.com/mirsayedshah/230704.git
df = pd.read_csv('feeder1_load_data.csv')

# Step 0: Sensitivity analysis
# Calculate mean and standard deviation for each bus.
mean_loads = df.mean()
std_devs = df.std()

# Calculate normalized standard deviation for each bus.
norm_std_devs = std_devs / mean_loads

# Print out the buses with the highest normalized standard deviations.
These are the buses that are most sensitive to load changes over time.
most_sensitive_buses = norm_std_devs.sort_values(ascending=False)
print(most_sensitive_buses)

# Step 1: Data Preparation
# No specific data preparation steps are considered for this case study.

# Step 2: Correlation Analysis
# Select numeric columns from the dataframe.
numeric_columns = df.select_dtypes(include=[float, int]).columns
# Calculate correlation matrix.
correlation_matrix = df[numeric_columns].corr()

# Visualize correlation matrix using a heatmap.
plt.figure(figsize=(10, 8))
sns.heatmap(correlation_matrix, annot=True, cmap='coolwarm')
```

```
plt.title('Correlation Matrix')
plt.savefig('correlation_matrix.png')
plt.show()
plt.close()

# Export correlation matrix to CSV file.
correlation_matrix.to_csv('correlation_matrix.csv', index=True)

# Step 3: Regression Analysis
# Prepare the target variable and explanatory variables.
X = df[['B01', 'B02', 'B04', 'B05', 'B06', 'B07', 'B08', 'B09', 'B10',
'B11', 'B12', 'B13', 'B14', 'B15']]
y = df['B03']

# Add a constant term to the explanatory variables.
X = sm.add_constant(X)

# Fit the regression model.
model = sm.OLS(y, X)
results = model.fit()

# Print regression results summary.
print(results.summary())

# Export regression results to CSV file.
results_summary = results.summary().tables[1].as_csv()
with open('regression_results.csv', 'w') as f:
    f.write(results_summary)

# Visualization of regression results.
plt.figure(figsize=(10, 6))
plt.scatter(y, results.fittedvalues)
plt.xlabel('Actual B03')
plt.ylabel('Predicted B03')
plt.title('Regression Results - Actual vs Predicted B03')
plt.savefig('regression_results.png')
plt.show()
plt.close()

# Create a DataFrame with actual and predicted values.
regression_results_df = pd.DataFrame({
    'Actual B03': y,
    'Predicted B03': results.fittedvalues
})

# Export regression results (actual vs predicted B03) to CSV.
regression_results_df.to_csv('regression_results_actual_vs_predicted.c
sv', index=False)

# Step 4: Granger Causality Analysis
# Perform the Granger causality test.
with warnings.catch_warnings():
    warnings.simplefilter("ignore", category=FutureWarning)
    granger_results = sm.tsa.stattools.grangercausalitytests(df[['B03',
'B01']], maxlag=3, verbose=False)

# Export Granger causality results to CSV file.
granger_results_summary = ""
```

```
for lag in granger_results.keys():
    result = granger_results[lag][0]['ssr_chi2test']
    result_summary = f"Lag: {lag}\n"
    result_summary += f"F-statistic: {result[0]}\n"
    result_summary += f"p-value: {result[1]}\n\n"
    granger_results_summary += result_summary

with open('granger_causality_results.csv', 'w') as f:
    f.write(granger_results_summary)

# Print Granger causality results.
for lag in granger_results.keys():
    print(f'Lag: {lag}')
    print(granger_results[lag])

# Visualization of Granger causality results.
lag_values = []
p_values = []

for lag in granger_results.keys():
    lag_values.append(lag)
    p_values.append(granger_results[lag][0]['ssr_chi2test'][1])

plt.figure(figsize=(10, 6))
plt.plot(lag_values, p_values, marker='o')
plt.xlabel('Lag')
plt.ylabel('p-value')
plt.title('Granger Causality Results')
plt.xticks(lag_values)
plt.axhline(y=0.05, color='red', linestyle='--', label='Significance
Threshold (p=0.05)')
plt.legend()
plt.savefig('granger_causality_results.png')
plt.show()
plt.close()


# This script is designed to be flexible and adaptable. Additional
visualizations, tables, calculations, and summaries can be incorporated
as needed to present the results. These elements can be tailored and
modified based on specific requirements, ensuring a comprehensive and
customized data analysis outcome.

# ...

# End of the script

"""
Author: Mir Sayed Shah Danish
Date: July 4, 2023 (Last updated)

This script was developed for educational and informational purposes
only.

Acknowledgments:
The author expresses his sincere gratitude to those who have added value
through their insightful feedback and significant contributions to this
project.
For any questions or further information, please contact
danish@mdanish.me.
"""
```

References

1. Zhang, Z., Yang, H., Yin, X., Han, J., Wang, Y., Chen, G.: A load-shedding model based on sensitivity analysis in on-line power system operation risk assessment. Energies. **11**, 727 (2018). https://doi.org/10.3390/en11040727
2. Danish, M.S.S.: Voltage Stability in Electric Power System: a Practical Introduction. Logos Verlag Berlin GmbH, Berlin (2015)
3. Vuluvala, M.R., Saini, L.M.: Load balancing of electrical power distribution system: an overview. In: 2018 International Conference on Power, Instrumentation, Control and Computing (PICC), pp. 1–5. IEEE, Thrissur (2018). https://doi.org/10.1109/PICC.2018.8384780
4. Furukakoi, M., Adewuyi, O.B., Danish, M.S.S., Howlader, A.M., Senjyu, T., Funabashi, T.: Critical boundary index (CBI) based on active and reactive power deviations. Int. J. Electr. Power Energy Syst. **100**, 50–57 (2018). https://doi.org/10.1016/j.ijepes.2018.02.010
5. Maaruf, M., Khan, K., Khalid, M.: Robust control for optimized islanded and grid-connected operation of solar/wind/battery hybrid energy. Sustain. For. **14**, 5673 (2022). https://doi.org/10.3390/su14095673
6. Brucoli, M., Maione, B., Margarita, E., Torelli, F.: Sensitivity analysis in power system dynamic stability studies. Electr. Power Syst. Res. **4**, 59–66 (1981). https://doi.org/10.1016/0378-7796(81)90037-7
7. Amusan, O.T., Nwulu, N.I., Gbadamosi, S.L.: Identification of weak buses for optimal load shedding using differential evolution. Sustain. For. **14**, 3146 (2022). https://doi.org/10.3390/su14063146
8. Chureemart, J., Churueang, P.: Sensitivity analysis and its applications in power system improvements. In: 2008 5th International Conference on Electrical Engineering/Electronics, Computer, Telecommunications and Information Technology, pp. 945–948 (2008). https://doi.org/10.1109/ECTICON.2008.4600587
9. Zhou, Y., Zhu, H.: Bus split sensitivity analysis for enhanced security in power system operations. In: 2019 North American Power Symposium (NAPS), pp. 1–6 (2019), https://doi.org/10.1109/NAPS46351.2019.9000243
10. Danish, M.S.S., Yona, A., Senjyu, T.: A review of voltage stability assessment techniques with an improved voltage stability indicator. Int. J. Emerg. Electr. Power Syst. **16**, 107–115 (2015). https://doi.org/10.1515/ijeeps-2014-0167
11. Ahmadi, M., Danish, M.S.S., Senjyu, T., Fedayee, H., Sabory, N.R., Yona, A.: Optimal merging of transportation system using renewable energy-based supply for sustainable development. In: Danish, M.S.S., Senjyu, T., Sabory, N.R. (eds.) Sustainability Outreach in Developing Countries, pp. 47–63. Springer, Singapore (2021). https://doi.org/10.1007/978-981-15-7179-4_4
12. Danish, M.S.S., Nazari, Z., Senjyu, T.: AI-coherent data-driven forecasting model for a combined cycle power plant. Energy Convers. Manag. **286**, 117063 (2023). https://doi.org/10.1016/j.enconman.2023.117063
13. Bu, F., Yuan, Y., Wang, Z., Dehghanpour, K., Kimber, A.: A time-series distribution test system based on real utility data. In: 2019 North American Power Symposium (NAPS), pp. 1–6. Iowa State University, Ames (2019). https://doi.org/10.1109/NAPS46351.2019.8999982
14. Danish, M.S.S., Senjyu, T., Danish, S.M.S., Sabory, N.R., Narayanan, K., Mandal, P.: A recap of voltage stability indices in the past three decades. Energies. **12**, 1544 (2019). https://doi.org/10.3390/en12081544
15. Saltelli, A., Ratto, M., Andres, T., Campolongo, F., Cariboni, J., Gatelli, D., Saisana, M., Tarantola, S.: Global Sensitivity Analysis: the Primer. Wiley-Interscience, Chichester/Hoboken (2008)
16. Ahmadi, M., Adewuyi, O.B., Danish, M.S.S., Mandal, P., Yona, A., Senjyu, T.: Optimum coordination of centralized and distributed renewable power generation incorporating battery storage system into the electric distribution network. Int. J. Electr. Power Energy Syst. **125**, 106458 (2021). https://doi.org/10.1016/j.ijepes.2020.106458

17. Zimmerman, R.D., Murillo-Sánchez, C.E., Thomas, R.J.: MATPOWER: steady-state operations, planning, and analysis tools for power systems research and education. IEEE Trans. Power Syst. **26**, 12–19 (2011). https://doi.org/10.1109/TPWRS.2010.2051168
18. Danish, M.S.S.: A framework for modeling and optimization of data-driven energy systems using machine learning. IEEE Trans. Artif. Intell. **1–10** (2023). https://doi.org/10.1109/TAI.2023.3322395
19. Danish, M.S.S.: AI and expert insights for sustainable energy future. Energies. **16**, 3309 (2023). https://doi.org/10.3390/en16083309
20. Danish, M.S.S.: AI in energy: overcoming unforeseen obstacles. AI. **4**, 406–425 (2023). https://doi.org/10.3390/ai4020022
21. Danish, M.S.S., Sabory, N.R., Funabashi, T., Danish, S.M.S., Noorzad, A.S., Yona, A., Senjyu, T.: Comparative analysis of load flow calculation methods with considering the voltage stability constraints. In: 2016 IEEE International Conference on Power and Energy (PECon), pp. 250–255 (2016), https://doi.org/10.1109/PECON.2016.7951568
22. Khuong, N.V., Shabbir, M.S., Sial, M.S., Khanh, T.H.T.: Does informal economy impede economic growth? Evidence from an emerging economy. J. Sustain. Finance Invest. **11**, 103–122 (2021). https://doi.org/10.1080/20430795.2020.1711501
23. Yang, F., Xiao, D.: Progress in root cause and fault propagation analysis of large-scale industrial processes. J. Control Sci. Eng. **2012**, e478373 (2012). https://doi.org/10.1155/2012/478373
24. Kumbhar, A., Dhawale, P.G., Kumbhar, S., Patil, U., Magdum, P.: A comprehensive review: machine learning and its application in integrated power system. Energy Rep. **7**, 5467–5474 (2021). https://doi.org/10.1016/j.egyr.2021.08.133
25. Poudel, S., Dubey, A., Schneider, K.P.: A generalized framework for service restoration in a resilient power distribution system. IEEE Syst. J. **16**, 252–263 (2022). https://doi.org/10.1109/JSYST.2020.3011901
26. Shumway, R.H., Stoffer, D.S.: Time Series Analysis and Its Applications: with R Examples. Springer, New York (2017)
27. Scargle, J.D.: Studies in astronomical time series analysis. II. Statistical aspects of spectral analysis of unevenly spaced data. Astrophys. J. **263**, 835–853 (1982). https://doi.org/10.1086/160554
28. Martínez Torres, J., Pastor Pérez, J., Sancho Val, J., McNabola, A., Martínez Comesaña, M., Gallagher, J.: A functional data analysis approach for the detection of air pollution episodes and outliers: a case study in Dublin, Ireland. Mathematics. **8**, 225 (2020). https://doi.org/10.3390/math8020225
29. Hastie, T., Tibshirani, R., Friedman, J.: The Elements of Statistical Learning: Data Mining, Inference, and Prediction, 2nd edn. Springer, New York (2016)

An Overview of the Roles of Inverters and Converters in Microgrids

Alexey Mikhaylov

1 Introduction

Microgrids represent a paradigm shift in energy distribution, offering a more decentralized, efficient, and sustainable approach compared to traditional power grids [1]. At the heart of microgrid functionality are power inverters and converters, which are essential for converting and managing electrical energy between various forms [2]. These devices enable the integration of diverse energy sources, such as solar, wind, and batteries, into microgrids, making them pivotal in the transition to renewable energy sources [3, 4]. The evolution of inverter and converter technology has been marked by significant advances in semiconductor materials, control strategies, and system design [5]. This progress has led to improved efficiency, reliability, and adaptability in diverse applications, particularly in renewable energy systems [6]. The transition from simple linear models to more complex software-driven designs has facilitated the integration of these devices into smart grids, enabling dynamic energy management and real-time response to changing load conditions [7].

The design of inverters and converters for microgrids involves a myriad of considerations, including efficiency, reliability, cost-effectiveness, and compliance with regulatory standards [8]. Design methodologies have evolved to address these challenges, incorporating advanced techniques such as pulse width modulation (PWM) and innovative components like LCL filters [9, 10]. These enhancements are crucial to minimizing harmonic distortion, optimizing power quality, and ensuring grid compatibility [11]. The advancements in inverter and converter technology have had a profound impact on the integration of renewable energy sources into a power grid [12]. Improved inverter designs, characterized by higher efficiency and better control over power quality, have made it feasible to harness solar and wind energy on a larger scale. These technologies play a vital role in enhancing the resilience and sustainability of power systems, aligning with global efforts to combat climate change and reduce dependence on fossil fuels [13, 14]. As the demand for

A. Mikhaylov (✉)
Research Center of Monetary Relations, Financial University Under the Government of the Russian, Moscow, Russia

M. S. S. Danish (ed.), *Unified Vision for a Sustainable Future*,
https://doi.org/10.1007/978-3-031-53574-1_3

renewable energy sources increases [15], the development of more advanced inverters and converters becomes crucial. Future trends in this domain include the incorporation of artificial intelligence and machine learning for predictive maintenance, enhanced grid-support functionalities, and the development of more compact, cost-effective designs [16, 17]. However, these advancements come with challenges such as the need for standardized protocols, interoperability issues, and ensuring cybersecurity in increasingly connected energy systems [18].

This study aims to provide a comprehensive overview of the roles of inverters and converters in microgrids, highlighting their importance in modern power systems. It delves into the technical aspects of these devices, including design methodologies, performance optimization strategies, and the implications of recent technological advancements. This chapter also discusses the challenges and future prospects in the field, contributing to the ongoing research and development efforts in sustainable power systems [19–21]. In fact, this introduction sets the stage for a detailed exploration of inverter and converter technologies in microgrids. It underscores their critical role, technological advancements, design considerations, and future trends, providing a foundation for understanding their impact on renewable energy integration and sustainable power distribution.

2 Power Conversion in Microgrids

This subsection introduces the concept of power conversion within the microgrid context. It outlines the fundamental need for power conversion in microgrids, which often combine various types of energy sources, such as solar, wind, and traditional generators, with different electrical characteristics [1]. This section explains how power converters and inverters bridge the gap between these diverse energy sources and the electrical loads or grid requirements, emphasizing their roles in maintaining stability, efficiency, and reliability in microgrids [22]. Also, it discusses the different types of converters and inverters commonly employed in microgrid applications. This does not cover all types of inverters and converters such as AC-DC, DC-AC, DC-DC, and AC-AC converters, which serve specific roles such as converting solar energy into usable AC power or adjusting DC voltage levels from storage batteries [23].

This study also deals with the technological aspects of power conversion, focusing on critical components like semiconductor devices, transformers, filters, and control systems. It explains how advancements in semiconductor technology, such as the development of IGBTs and MOSFETs, have enhanced the performance of power converters. The role of filters (like LCL filters) in minimizing harmonic distortion and the importance of transformers for voltage adaptation and isolation are also discussed. Effective control strategies are crucial for the efficient operation of power converters in microgrids [24]. This section examines various control methods, including pulse width modulation (PWM), maximum power point tracking (MPPT) for solar inverters, and control algorithms for maintaining voltage and frequency stability [25]. This study highlights the importance of these strategies in optimizing the conversion process, ensuring power quality, and facilitating the integration of renewable sources [26–28].

Power conversion in microgrids faces unique challenges such as managing variable renewable energy sources, ensuring compatibility with the main grid, dealing with load fluctuations, and maintaining reliability in isolated or islanded microgrid operations [5, 13, 29]. The impact of these challenges on the design and selection of power converters is discussed along with potential solutions and ongoing research areas. Finally, this study summarizes the key points discussed, reinforcing the integral role of power conversion in the efficient operation of microgrids. It also provides a glimpse into the future of power conversion technology, touching upon emerging trends, potential advancements, and the evolving landscape of microgrid applications.

3 Power Converter/Inverter

A schematic representation of the stages of power conversion in an inverter involves converting solar energy into utility-scale electrical energy suitable for distribution via a power grid. Here is an explanation of each step illustrated in Fig. 1 [24, 30–35]:

- Step 1: Solar Panels
  Solar panels, composed of many solar cells, capture sunlight and convert it into direct current (DC). This is the first stage of the power conversion process, where photovoltaic (PV) effect is utilized.
- Step 2: DC Power Processing
  The DC electricity generated by the solar panels may require voltage level adjustments or regulation. This step likely involves DC-DC conversion to ensure that the DC output is at the proper voltage level for either storage or inversion. This stage may include components like charge controllers.
- Step 3: Inverter (DC to AC Conversion)
  The DC power is then fed into an inverter. Here, the DC is converted into alternating current (AC). This is a crucial step for integrating solar-generated electric-

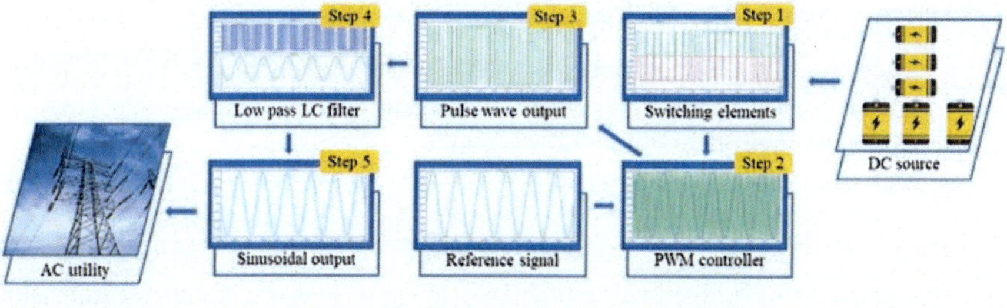

Fig. 1 Block diagram of the DC to AC conversion process in an inverter

ity into a power grid, which operates on AC. The inverter may use techniques like pulse width modulation (PWM) to convert and control the output waveform quality.

- Step 4: AC Power Processing
  Once the electricity is in AC form, it may need further processing to match the grid's frequency and voltage requirements. This involves filtering to smooth out the waveform and transformers to step up the voltage to the appropriate level for grid distribution.
- Step 5: Transmission Grid
  The final step involves transmitting the processed AC electricity into the power grid. This stage symbolizes the integration of solar-generated electricity into the broader electrical grid, allowing it to be distributed to homes, businesses, and other end-users.

3.1 Design Approach and Methodology

One of the key techniques enhancing the performance of these inverters is pulse width modulation (PWM), particularly the sine-triangle space-vector modulation method [24]. This technique stands out for its ability to reduce harmonic distortion and optimize the efficiency of voltage source inverters. It plays a pivotal role in managing the output voltage quality, which is a critical aspect of inverter functionality. Another integral component of inverter design is an LCL filter [10]. This filter, which is crucial for mitigating harmonics and ensuring power quality, must be carefully designed to align with the specific requirements of the system. Recent advancements and research in LCL filter design have focused on optimizing these components for various applications, particularly in grid-connected scenarios, to enhance the overall efficiency and reliability of the inverter system.

Together, these components and techniques form the backbone of modern inverter design, contributing significantly to the advancement and integration of renewable energy sources into the existing power grid, thereby shaping the future of sustainable power generation and distribution [36]. According to the literature [2, 8–10, 24, 30, 37–40], the design and methodology steps of an inverter/converter can be summarized as follows:

- System Requirements Analysis
  - Thoroughly review and document all necessary specifications such as voltage levels, power capacity, efficiency, and operational frequencies.
  - Consider environmental and operational conditions under which the inverter will function.
- Topology Selection
  - Opt for a full-bridge inverter topology using insulated gate bipolar transistors (IGBTs) for their high efficiency and fast switching capabilities.

    - Analyze the benefits and limitations of this topology in the context of system's requirements.

- Pulse Width Modulation (PWM) Technique
    - Implement sine-triangle space-vector modulation to optimize the efficiency and minimize the harmonic distortion in the output.
    - Design the control logic for PWM to ensure precise switching of IGBTs.
- Transformer Design
    - Design a high-voltage transformer focusing on core material selection, turns ratio, and leakage inductance calculations.
    - Ensure the transformer's capability to handle the 10 kVA power rating and provide the necessary electrical isolation.
- Filter Design
    - Design an LCL filter to effectively mitigate harmonics and output voltage distortion.
    - Calculate the optimal values for the inductors and capacitors, tailored for the 60 Hz output frequency.
- Feedback Control System
    - Develop PID controllers for voltage and current regulation.
    - Systematically tune the PID parameters to achieve stability and the desired dynamic response.
- Protection Mechanisms
    - Incorporate protection features against overvoltage, overcurrent, short-circuit, and overtemperature scenarios.
    - Design these mechanisms to swiftly and safely shut down or isolate the inverter when necessary.
- Thermal Management
    - Design an efficient cooling system, choosing between forced air, natural convection, or liquid cooling based on thermal analysis.
    - Ensure that the cooling solution aligns with the inverter's power dissipation and environmental conditions.
- Power Factor Correction
    - Implement a boost PFC topology to optimize the input power factor, leading to improved energy efficiency.
- Grid Synchronization
    - Utilize the phase-locked loop (PLL) or zero-crossing detection techniques for precise synchronization of the inverter output with the grid frequency and phase.

- Anti-islanding Protection
  - Implement both passive and active anti-islanding techniques to ensure safe operation and compliance with grid interconnection standards.
- Regulatory Compliance
  - Thoroughly research and ensure adherence to relevant IEC, IEEE, and local standards governing the inverter design and grid interconnection.
- Enclosure Design
  - Design an enclosure considering factors like ingress protection (IP) rating, mechanical strength, ease of installation, and thermal management.
- Safety and Reliability Engineering
  - Integrate component derating, redundancy, and fault-tolerant design principles to enhance safety and reliability.
- User Interface and Communication
  - Develop an intuitive user interface for monitoring, diagnostics, and control, considering options for both local and remote access.
- Cost and Size Optimization
  - Throughout the design process, balance performance with cost and size constraints to meet project budgets and physical space limitations.
- MATLAB Simulink Modeling
  - Build a comprehensive model of the entire inverter system in MATLAB Simulink using suitable libraries and blocks.
  - Run simulations to evaluate performance against specifications and adjust the design accordingly.
- Prototype Testing and Validation
  - After successful simulation, construct a prototype and perform rigorous testing to validate the design under real-world conditions.
  - Document the test results and compare them against the theoretical performance to ensure compliance with all specifications.
- Iterative Design Improvement
  - Based on testing feedback, iterate the design to resolve any issues or to enhance performance.
- Final Design Documentation
  - Prepare detailed design document including schematics, PCB layouts, software code, and user manuals.
- Manufacturing and Assembly Planning
  - Plan for the manufacturing process, considering aspects like component sourcing, assembly procedures, and quality control measures.Market and Environmental ConsiderationsAnalyze the market viability and environmental impact of the inverter, considering sustainability and end-of-life disposal.

3.2 Design Parameters and Specifications

In the quest to design an efficient and robust voltage inverter for microgrid applications, it is imperative to meticulously define and adhere to a set of design parameters and specifications. These parameters form the blueprint for the inverter's development, ensuring that the final product meets both performance requirements and regulatory standards [8]. Table 1 presents a detailed breakdown of these design parameters, outlining the specific technical specifications and additional remarks that guide the design process.

Table 1 provides a comprehensive reference point, addressing various aspects of the inverter design ranging from electrical input and output characteristics to more nuanced elements like cooling systems and safety features. Each parameter is carefully chosen to align with the overarching goal of creating an inverter that is not only efficient and reliable but also compliant with current technological and environmental standards [9, 37].

The design and modeling of an inverter require some criteria to meet the techno-economic standards, for example, fast switching techniques, wider operating temperature range, high electrical and thermal efficiency, fast fault detection and isolation, and long life span [36].

The pulse width modulation (PWM) technique [9, 39] in inverter technology is a fundamental method for controlling power delivery to the load (Fig. 2). In this method, a gating signal, typically generated by a control unit, is used to operate the switching devices in a manner that varies the pulse width, or the duty cycle, of the output voltage in proportion to the input voltage. This technique allows the modulation of the output voltage while maintaining control over the power delivered.

In the context of a 60 Hz output frequency, which is standard in many power systems, the PWM technique entails flipping the switching frequency every 8.333 milliseconds (ms). This corresponds to the periodicity of a 60 Hz signal, where one cycle lasts for approximately 16.667 ms. Therefore, to achieve this frequency, the switches change their state (either on or off) at the halfway point of each cycle, which is at 8.333 ms intervals. The operation of the switches follows a specific sequence to ensure the correct formation of the output voltage waveform. For instance, in a full-bridge inverter configuration, switches S1 and S4 are turned on simultaneously, while S2 and S3 are turned off. After a predetermined period, the states of these switches are reversed – S1 and S4 are turned off, while S2 and S3 are turned on. This alternating pattern of switching creates a waveform that can be modulated to resemble the desired AC output.

The effectiveness of this technique in creating a sinusoidal output lies in the ability to adjust the duty cycle of the modulated pulses. The duty cycle, which is the ratio of the "on" time to the total cycle time, is proportional to the amplitude of the sinusoidal signal being encoded. By varying this duty cycle over time, the inverter can effectively recreate the sinusoidal waveform at the desired frequency and amplitude.

The PWM technique, particularly when implemented in inverter designs using advanced methods like sine-triangle space-vector modulation, offers a highly

Table 1 Design parameters and specifications of a microgrid voltage inverter

Design parameter	Specification	Remarks
Input voltage	400 V	DC
Output voltage	$6600\frac{\sqrt{3}}{\sqrt{2}}$ V (5388.877434 V)	AC
Output frequency	60 Hz	
Phase	Single phase	
Power rating	10 kVA	
Efficiency	80–95%	
Topology	Full bridge	
Switching devices	IGBTs	
Switching frequency	Based on losses, EMI, and size constraints	
PWM techniques	Sine triangle	Space-vector modulation
Isolation	High-voltage transformer	
Transformer design	Core material, turns ratio, leakage inductance, etc.	
Filter design	LCL	
Feedback control	PID	Voltage and current control algorithms
Protection features	Overvoltage, overcurrent, short-circuit, and over-temperature	
Cooling system	Forced air, natural convection, or liquid cooling	
Power factor correction (PFC)	Boost	Buck-boost or another suitable topology
Grid synchronization	Phase-locked loop (PLL), zero-crossing detection, or other techniques	
Anti-islanding protection	Passive or active methods	
Regulatory standards	IEC	Compliance with relevant standards (e.g., IEEE)
Enclosure design	IP rating, mechanical strength, and thermal management	
Safety and reliability	Component derating, redundancy, and fault-tolerant design	
User interface	Local or remote monitoring, diagnostics, and control	
Cost and size constraints	Balancing performance with budget and space limitations	

efficient way to control the output voltage and frequency, which is crucial in applications like microgrids and renewable energy systems [39].

The reference to a "50% duty cycle" in this context is particularly significant. The duty cycle in PWM refers to the proportion of one cycle in which a signal or power switch is active (or "on"). A 50% duty cycle means that the switch is on for half of the cycle time and off for the other half. In the context of sine-wave

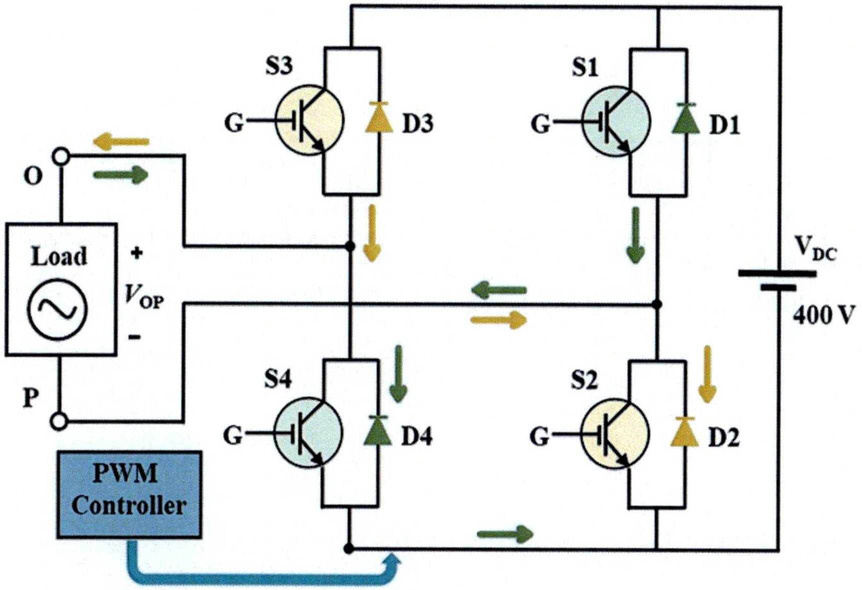

Fig. 2 Schematic model of the inverter

formation, this 50% duty cycle is used as a baseline or a reference point around which the actual modulated waveform is created (Figs. 3, 4, and 5).

To achieve a sinusoidal output, the PWM control strategy varies the duty cycle of the switching signals throughout each cycle of the output waveform. This variation is not random but follows the shape of the desired sine wave. At the start and end of the sine-wave cycle (which corresponds to the zero-crossing points of the sine wave), the duty cycle is near 50%, meaning the switch is on for approximately half the time. As the sine wave reaches its peak or trough, the duty cycle increases or decreases correspondingly, allowing more or less power to flow through, which in turn shapes the output waveform to resemble a sine wave. The prediction or creation of this sine wave using PWM involves intricate control algorithms. These algorithms calculate the required duty cycle at each point in the cycle to replicate the sine wave's natural rise and fall. This method is particularly effective in inverter design, allowing the generation of high-quality AC power from a DC source.

In the domain of power electronics, particularly in inverter design, bi-polar sinusoidal pulse width modulation (SPWM) is a pivotal technique for controlling the output voltage to achieve a sinusoidal waveform. This method hinges on three fundamental components [41]: the reference voltage, the carrier voltage, and the control logic. The reference voltage is essentially the desired output waveform, typically a sinusoid, which the inverter aims to replicate. The carrier voltage, usually a high-frequency triangular wave, serves as a basis for modulating the reference signal. The core of this technique lies in the control logic, typically implemented as a comparator. This comparator receives both the reference and carrier voltages as inputs and generates a PWM signal as its output. The output signal's duty cycle is

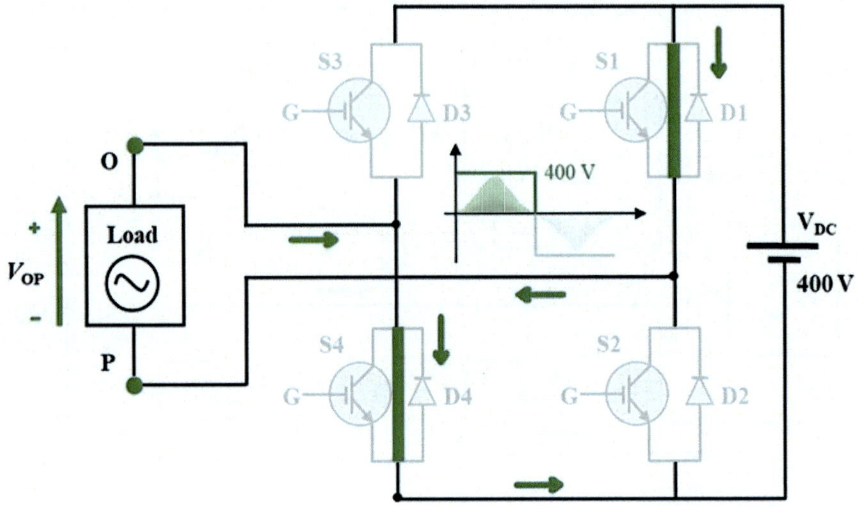

Fig. 3 Switching status of functioning switches 1 and 4

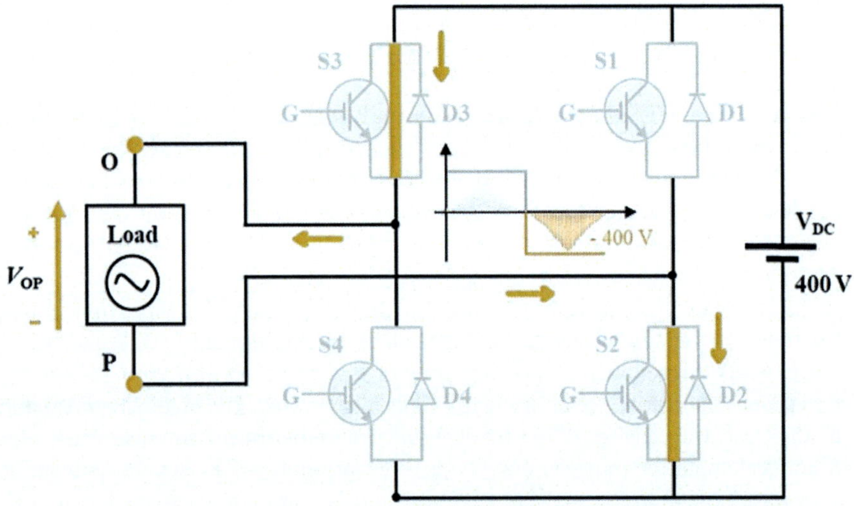

Fig. 4 Switching status of functioning switches 2 and 3

modulated based on the instantaneous comparison between the reference and the carrier signals (Figs. 6 and 7). When the reference voltage exceeds the carrier voltage, the comparator output is high, and vice versa. This mechanism results in a PWM output that varies in pulse width proportional to the reference sinusoidal voltage, thereby controlling the inverter output to closely mimic a sinusoidal waveform, as illustrated in Figs. 6 and 7. This approach is integral to achieving efficient,

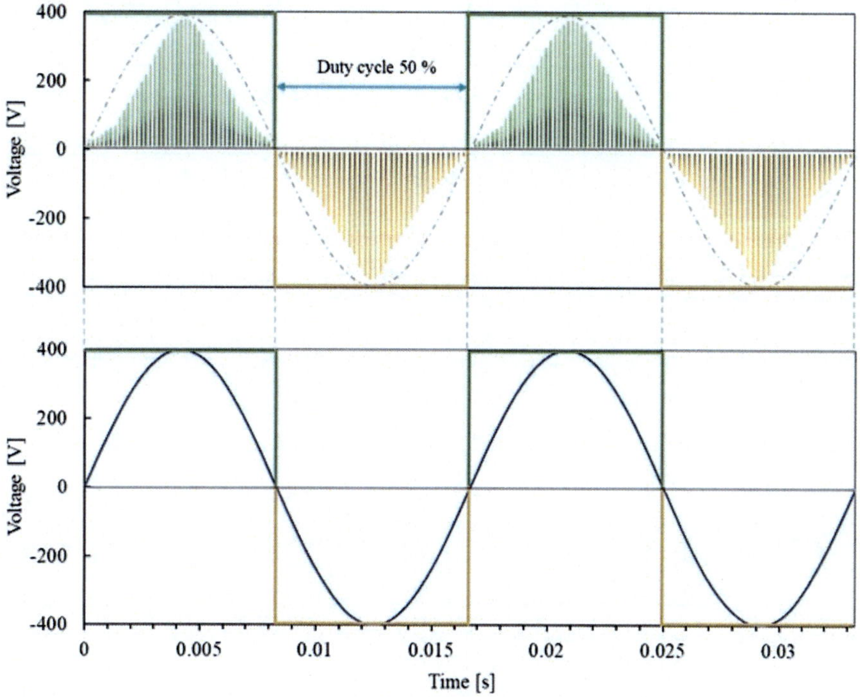

Fig. 5 Inverter switching pattern output before SWPM technique application

Fig. 6 PWM control block input and output relations

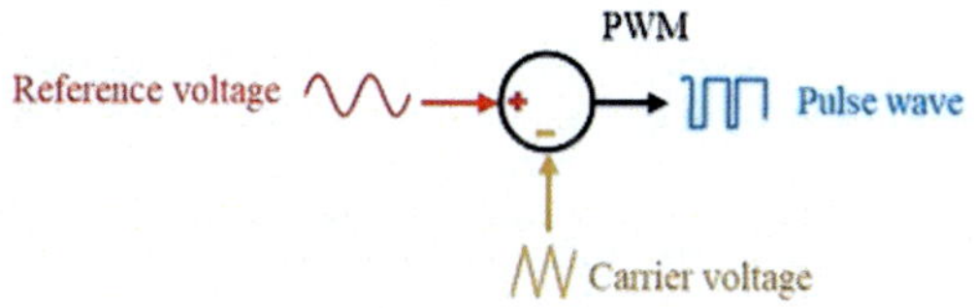

reliable, and high-quality AC power output in various applications, from grid-tied inverters to renewable energy systems.

Low-pass LC filter is designed based on the two factors switching ripple current ($\Delta_{\text{ripple, max}}$) and voltage drop (less than 3% of the rated or connected load) across the filter inductor, which does not have a considerable impact [42, 43]:

$$\Delta_{\text{ripple,max}} = \frac{1}{8}\frac{V_{DC}}{L f_{sw}} \Rightarrow L = \frac{1}{8}\frac{V_{DC}}{\Delta_{\text{ripple,max}} f_{sw}} \tag{1}$$

$$C = \frac{\alpha P_{\text{rated}}}{2\pi f_{sys} {V_{\text{rated}}}^2} \tag{2}$$

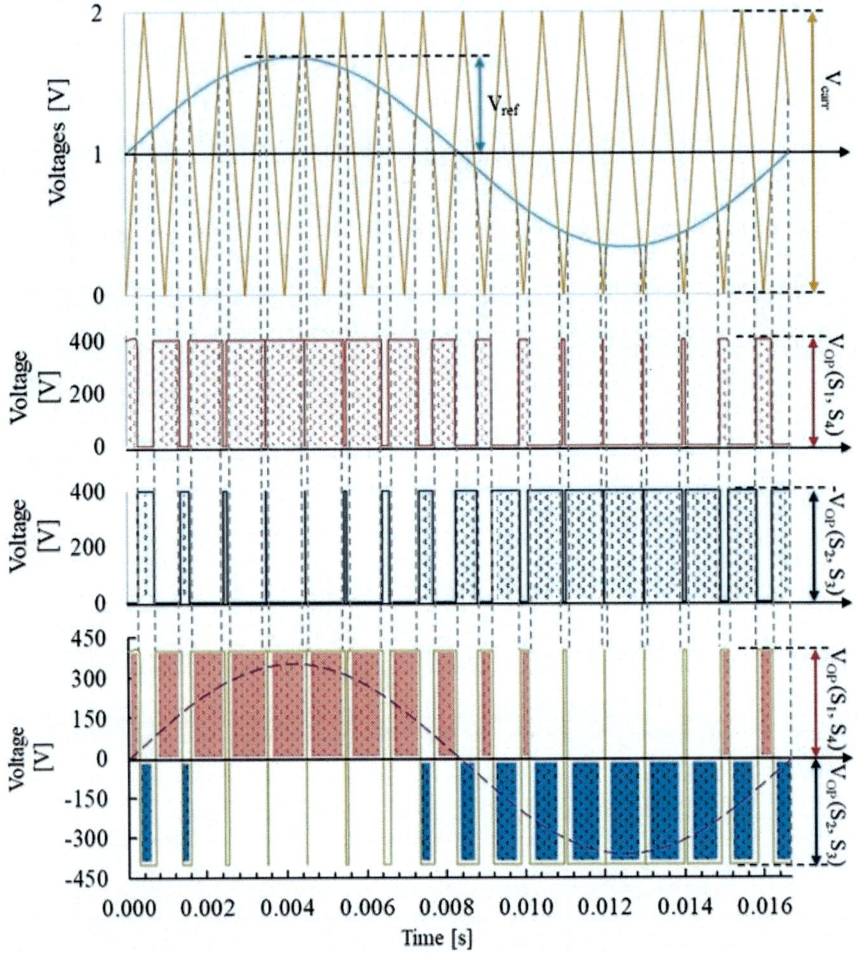

Fig. 7 Simulation of the SWPM control output based on the supposed reference and carrier voltage ratio

V_{DC}	Inverter input voltage [V]	400 V (DC)
V_{rated}	Rated voltage [V]	440 V (AC)
P_{rated}	Rated power [kW]	16.057 kW (obtained by ignoring α)
L	Low-pass filter series inductor [H]	3 mH
C	Low-pass filter series capacitor [F]	220 μF
f_{sys}	System frequency [Hz]	60 Hz
f_{sw}	Switching frequency [Hz]	1 kHz ~ 30 kHz
$\Delta_{ripple,\ max}$	Maximum ripple current [%]	16.7% at 1 kHz (recommended between 15% and 25%)
α	Reactive power factor [%]	5%

It is recommended to set the switching ripple current ($\Delta_{ripple,\ max}$) passing through the filter inductor to the consumer (AC side) between 15% and 25% of the peak magnitude of AC output current [32]. The switching ripple current and reactive power factor are supposed to be 20 and 5%, respectively [42]. Since this is a grid/network-tied inverter, the actual rating is not considered and calculated based on best performances in reference to the proposed parameters of the filter, ensuring the output voltage control rating. Filter parameter values can be changed by inductor value corrections if the filter is connected to a rotating load (motor) that can compensate for the connected internal inductance. .

4 Discussion and Results

This study underscores a paradigm shift in energy distribution, with microgrids representing a decentralized, efficient, and sustainable approach compared to traditional power grids. At the core of this transformation are power inverters and converters, which are essential for managing and converting electrical energy in various forms. These devices enable the integration of diverse energy sources such as solar, wind, and batteries into microgrids, making them pivotal in the transition to renewable energy sources. The evolution of inverter and converter technology is marked by significant advances in semiconductor materials, control strategies, and system design. This progress has led to improved efficiency, reliability, and adaptability in diverse applications, particularly in renewable energy systems. The transition from simple linear models to more complex software-driven designs has facilitated the integration of these devices into smart grids, enabling dynamic energy management and real-time response to changing load conditions.

A key technique enhancing the performance of inverters is pulse width modulation (PWM), particularly the sine-triangle space-vector modulation method. This technique is vital for reducing harmonic distortion and optimizing the efficiency of voltage source inverters. Moreover, an LCL filter, which is crucial for mitigating harmonics and ensuring power quality, must be carefully designed to align with the specific requirements of the system. These components form the backbone of modern inverter design, significantly contributing to the advancement and integration of renewable energy sources into the existing power grid. The integration of advanced inverters and converters in microgrids involves numerous challenges such as standardized protocols, interoperability issues, and ensuring cybersecurity in increasingly connected energy systems. Future trends include incorporating artificial intelligence and machine learning for predictive maintenance, enhanced grid-support functionalities, and the development of more compact, cost-effective designs. This study provides a comprehensive overview of the roles of inverters and converters in microgrids, highlighting their importance in modern power systems. It delves into the technical aspects of these devices, including design methodologies, performance optimization strategies, and the implications of recent technological advancements. The challenges and future prospects in the field are also discussed, contributing to the ongoing research and development efforts in sustainable power systems.

5 Conclusion

This chapter has presented an exploration of inverter and converter technologies in microgrids, emphasizing their critical roles in the integration of renewable energy and sustainable power distribution. The advancements in semiconductor technology, innovative design approaches like PWM, and components such as LCL filters have significantly improved the performance of these crucial devices.

The study also highlights the design approach and methodology for inverters and converters, encompassing system requirements analysis, topology selection, transformer design, filter design, feedback control systems, and various other crucial aspects. These methodologies are integral to enhancing the resilience and sustainability of power systems, aligning with global efforts to combat climate change and reduce dependence on fossil fuels. In conclusion, inverters and converters are central to the efficient operation of microgrids. Their continuous evolution, driven by technological advancements and design innovations, is essential in meeting the growing demand for renewable energy sources and shaping the future of sustainable power generation and distribution. The challenges identified in this study provide a roadmap for future research and development, ensuring that microgrids continue to be a viable solution for modern energy needs.

References

1. Danish, M.S.S., Senjyu, T., Funabashia, T., Ahmadi, M., Ibrahimi, A.M., Ohta, R., Rashid Howlader, H.O., Zaheb, H., Sabory, N.R., Sediqi, M.M.: A sustainable microgrid: a sustainability and management-oriented approach. Energy Procedia. **159**, 160–167 (2019). https://doi.org/10.1016/j.egypro.2018.12.045
2. Kim, Y.-J., Kim, H.: Optimal design of LCL filter in grid-connected inverters. IET Power Electron. **12**, 1774–1782 (2019). https://doi.org/10.1049/iet-pel.2018.5518
3. Danish, M.S.S.: Green building efficiency and sustainability indicators. In: Senjyu, T.S. (ed.) Green Building Management and Smart Automation, pp. 128–145. IGI Global, Hershey (2020). https://doi.org/10.4018/978-1-5225-9754-4.ch006
4. Danish, S.M.S., Zaheb, H., Sabori, N.R., Karimy, H., Faiq, A.B., Fedayi, H., Senjyu, T.: The road ahead for municipal solid waste management in the 21st century: a novel-standardized simulated paradigm. In: IOP Conference Series: Earth and Environmental Science, p. 012009. IOP Publishing (2019). https://doi.org/10.1088/1755-1315/291/1/012009
5. Yaqobi, M.A., Matayoshi, H., Danish, M.S.S., Urasaki, N., Howlader, A.M., Senjyu, T.: Control and energy management strategy of standalone DC microgrid cluster using PV and battery storage for rural application. Int. J. Power Energy Res. **2**, 53–68 (2018). https://doi.org/10.22606/ijper.2018.24001

6. Danish, M.S.S.: Voltage Stability in Electric Power System: a Practical Introduction. Logos Verlag Berlin GmbH, Berlin (2015)
7. Danish, M.S.S., Nazari, Z., Senjyu, T.: AI-coherent data-driven forecasting model for a combined cycle power plant. Energy Convers. Manag. **286**, 117063 (2023). https://doi.org/10.1016/j.enconman.2023.117063
8. Wang, S., Xia, Z., Duan, H., Ma, C., Lu, S., Li, S.: A Si IGBT and SiC MOSFET hybrid full-bridge inverter and its modulation scheme. In: 2022 25th International Conference on Electrical Machines and Systems (ICEMS), pp. 1–5 (2022), https://doi.org/10.1109/ICEMS56177.2022.9983311
9. Bandaru, U., Priyadarshini, Y.S.I., Siva Sathyanarayana, M.: Space vector PWM algorithms for three-level inverter. Adv. Aspects Eng. Res. **16**, 30–51 (2021)
10. Khan, D., Qais, M., Sami, I., Hu, P., Zhu, K., Abdelaziz, A.Y.: Optimal LCL-filter design for a single-phase grid-connected inverter using metaheuristic algorithms. Comput. Electr. Eng. **110**, 108857 (2023). https://doi.org/10.1016/j.compeleceng.2023.108857
11. Danish, M.S.S., Senjyu, T., Ahmadi, M., Ludin, G.A., Ahadi, M.H., Karimy, H., Khosravy, M.: A review on energy efficiency for pathetic environmental trends mitigation. J. Sustain. Outreach. **2**(1–8), 1 (2021). https://doi.org/10.37357/1068/jso.2.1.01
12. Danish, M.S.S., Senjyu, T., Funabashi, T., Ahmadi, M., Ibrahimi, A.M., Ohta, R., Rashid Howlader, H.O., Zaheb, H., Sabory, N.R., Sediqi, M.M.: A sustainable microgrid: a sustainability and management-oriented approach. Energy Procedia. **159**, 160–167 (2019). https://doi.org/10.1016/j.egypro.2018.12.045
13. Danish, M.S.S., Matayoshi, H., Howlader, H.O.R., Chakraborty, S., Mandal, P., Senjyu, T.: Microgrid planning and design: resilience to sustainability. In: 2019 IEEE PES GTD Grand International Conference and Exposition Asia (GTD Asia), pp. 253–258. IEEE, Bangkok (2019). https://doi.org/10.1109/GTDAsia.2019.8716010
14. Danish, M.S.S., Senjyu, T.S. (eds.): System of green resilience eco-oriented land uses in urban socio-ecosystems. In: Eco-Friendly Energy Processes and Technologies for Achieving Sustainable Development, pp. 1–23. IGI Global, Hershey (2021). https://doi.org/10.4018/978-1-7998-4915-5
15. Furukakoi, M., Sediqi, M.M., Senjyu, T., Danish, M.S.S., Howlader, A.M., Hassan, M.A.M., Funabashi, T.: Optimum capacity of energy storage system considering solar radiation forecast error and demand response. In: 2017 IEEE 3rd International Future Energy Electronics Conference and ECCE Asia (IFEEC 2017 – ECCE Asia), pp. 997–1001. IEEE, Kaohsiung (2017). https://doi.org/10.1109/IFEEC.2017.7992177
16. Danish, M.S.S.: AI and expert insights for sustainable energy future. Energies. **16**, 3309 (2023). https://doi.org/10.3390/en16083309
17. Danish, M.S.S.: AI in energy: overcoming unforeseen obstacles. AI. **4**, 406–425 (2023). https://doi.org/10.3390/ai4020022
18. Umar, T., Egbu, C., Ofori, G., Honnurvali, M.S., Saidani, M., Opoku, A.: Challenges towards renewable energy: an exploratory study from the Arabian Gulf region. Proc. Inst. Civil Eng. Energy. **173**, 68–80 (2020). https://doi.org/10.1680/jener.19.00034
19. Ernst, D., Glavic, M., Wehenkel, L.: Power systems stability control: reinforcement learning framework. IEEE Trans. Power Syst. **19**, 427–435 (2004). https://doi.org/10.1109/TPWRS.2003.821457
20. Duan, C., Jiang, L., Fang, W., Liu, J.: Data-driven Affinely adjustable distributionally robust unit commitment. IEEE Trans. Power Syst. **33**, 1385–1398 (2018). https://doi.org/10.1109/TPWRS.2017.2741506
21. Khalid, H.M., Flitti, F., Mahmoud, M.S., Hamdan, M.M., Muyeen, S.M., Dong, Z.Y.: Wide area monitoring system operations in modern power grids: a median regression function-based state estimation approach towards cyber attacks. Sustain. Energy Grids Netw. **34**, 101009 (2023). https://doi.org/10.1016/j.segan.2023.101009
22. Danish, M.S.S., Matayoshi, H., Howlader, H.R., Chakraborty, S., Mandal, P., Senjyu, T.: Microgrid planning and design: resilience to sustainability. In: 2019 IEEE PES GTD Grand

International Conference and Exposition Asia (GTD Asia), pp. 253–258 (2019). https://doi.org/10.1109/GTDAsia.2019.8716010
23. Ahmadi, M., Adewuyi, O.B., Danish, M.S.S., Mandal, P., Yona, A., Senjyu, T.: Optimum coordination of centralized and distributed renewable power generation incorporating battery storage system into the electric distribution network. Int. J. Electr. Power Energy Syst. **125**, 106458 (2021). https://doi.org/10.1016/j.ijepes.2020.106458
24. Wang, F.: Sine-triangle versus space-vector modulation for three-level PWM voltage-source inverters. IEEE Trans. Ind. Appl. **38**, 500–506 (2002). https://doi.org/10.1109/28.993172
25. Singh, S.K., Lohani, B., Arora, L., Choudhary, D., Nagarajan, B.: A visual-inertial system to determine accurate solar insolation and optimal PV panel orientation at a point and over an area. Renew. Energy. **154**, 223–238 (2020). https://doi.org/10.1016/j.renene.2020.02.107
26. Sagara, M., Furukakoi, M., Senjyu, T., Danish, M.S.S., Funabashi, T.: Voltage stability improvement to power systems with energy storage systems. In: 2016 17th International Conference on Harmonics and Quality of Power (ICHQP), pp. 7–10. IEEE, Belo Horizonte (2016). https://doi.org/10.1109/ICHQP.2016.7783463
27. Danish, M.S.S., Senjyu, T., Danish, S.M.S., Sabory, N.R., Narayanan, K., Mandal, P.: A recap of voltage stability indices in the past three decades. Energies. **12**, 1544 (2019). https://doi.org/10.3390/en12081544
28. Zaheb, H., Danish, M.S.S., Senjyu, T., Ahmadi, M., Nazari, A.M., Wali, M., Khosravy, M., Mandal, P.: A contemporary novel classification of voltage stability indices. Appl. Sci. **10**, 1639 (2020). https://doi.org/10.3390/app10051639
29. Yaqobi, M.A., Matayoshi, H., Danish, M.S.S., Lotfy, M.E., Howlader, A.M., Tomonobu, S.: Low-voltage solid-state DC breaker for fault protection applications in isolated DC microgrid cluster. Appl. Sci. **9**, 723–735 (2019). https://doi.org/10.3390/app9040723
30. Bennia, I., Daili, Y., Harrag, A.: LCL filter design for low voltage-source inverter. In: Hatti, M. (ed.) Artificial Intelligence and Heuristics for Smart Energy Efficiency in Smart Cities, pp. 332–341. Springer International Publishing, Cham (2022). https://doi.org/10.1007/978-3-030-92038-8_34
31. Sward, J.A., Siff, J., Gu, J., Zhang, K.M.: Strategic planning for utility-scale solar photovoltaic development – Historical peak events revisited. Appl. Energy. **250**, 1292–1301 (2019). https://doi.org/10.1016/j.apenergy.2019.04.178
32. Mishra, P., Maheshwari, R.: Design, analysis, and impacts of sinusoidal LC filter on Pulsewidth modulated inverter fed-induction motor drive. IEEE Trans. Ind. Electron. **67**, 2678–2688 (2020). https://doi.org/10.1109/TIE.2019.2913824
33. Ahmadi, M., Danish, M.S.S., Lotfy, M.E., Yona, A., Hong, Y.-Y., Senjyu, T.: Multi-objective time-variant optimum automatic and fixed type of capacitor bank allocation considering minimization of switching steps. AIMS Energy. **7**, 792 (2019). https://doi.org/10.3934/energy.2019.6.792
34. Indu Rani, B., Saravana Ilango, G., Nagamani, C.: Power flow management algorithm for photovoltaic systems feeding DC/AC loads. Renew. Energy. **43**, 267–275 (2012). https://doi.org/10.1016/j.renene.2011.11.035
35. Danish, M.S.S., Sabory, N.R., Funabashi, T., Danish, S.M.S., Noorzad, A.S., Yona, A., Senjyu, T.: Comparative analysis of load flow calculation methods with considering the voltage stability constraints. In: 2016 IEEE International Conference on Power and Energy (PECon), pp. 250–255. IEEE, Melaka (2016). https://doi.org/10.1109/PECON.2016.7951568
36. Chapman, S.: Electric Machinery Fundamentals. McGraw-Hill Science/Engineering/Math, New York (2004)
37. Abdalgader, I.A.S., Kivrak, S., Özer, T.: Power performance comparison of SiC-IGBT and Si-IGBT switches in a three-phase inverter for aircraft applications. Micromachines (Basel). **13**, 313 (2022). https://doi.org/10.3390/mi13020313
38. Tang, Z., Sangwongwanich, A., Yang, Y., Blaabjerg, F.: Energy efficiency enhancement in full-bridge PV inverters with advanced modulations. E-prime – Advances in electrical engineering. Electron. Energy. **1** (2021). https://doi.org/10.1016/j.prime.2021.100004

39. Wang, F.: Sine-triangle vs. space vector modulation for three-level PWM voltage source inverters. In: Conference Record of the 2000 IEEE Industry Applications Conference. Thirty-Fifth IAS Annual Meeting and World Conference on Industrial Applications of Electrical Energy (Cat. No.00CH37129). 4, 2482–2488 (2000). https://doi.org/10.1109/IAS.2000.883171
40. Fan, Z., Yi, H., Xu, J., Xie, K.: Passivity-based design for LCL-filtered grid-connected inverters with inverter current control and capacitor-current active damping. J. Electr. Eng. Technol. (2023). https://doi.org/10.1007/s42835-023-01659-w
41. Koreboina, V.B., Narasimharaju, B.L., DM, V.K.: Performance evaluation of switched reluctance motor PWM control in PV-fed water pump system. Int. J. Renew. Energy Res. (IJRER). **6**, 941–950 (2016)
42. Azri, M., Rahim, N.A.: Design analysis of low-pass passive filter in single-phase grid-connected transformerless inverter. In: 2011 IEEE Conference on Clean Energy and Technology (CET), pp. 348–353. IEEE, Kuala Lumpur (2011). https://doi.org/10.1109/CET.2011.6041489
43. IGBT Technologies and Applications Overview: How and When to Use an IGBT. Semiconductor Components Industries, LLC, Phoenix (2018)

Integrating Machine Learning into Energy Systems: A Techno-economic Framework for Enhancing Grid Efficiency and Reliability

Mohammad Hamid Ahadi

1 Introduction

Energy management systems (EMS) are the backbone of modern electrical grids, playing a pivotal role in balancing supply with demand, ensuring reliability, and optimizing operations for improved efficiency [1]. As the world grapples with the imperative of sustainable energy use, the EMS have evolved from simple grid management tools to sophisticated systems capable of making real-time, data-driven decisions [2]. Historically, grid management was a relatively straightforward affair: match supply with demand, maintain system balance, and ensure a steady flow of electricity [3, 4]. However, the landscape of energy management has been transformed by the emergence of renewable energy sources, the decentralization of energy systems, and the digitalization of grid operations. The increasing unpredictability of supply from sources such as wind and solar power, the rising demand for electricity, and the pressing need for environmental conservation have all contributed to the complexity of modern energy systems [5].

The purpose of this chapter is to explore the complex interplay between the various elements of the modern EMS. By focusing on the crucial aspects of efficiency and reliability, this study aims to dissect the challenges and innovations shaping today's energy landscape. This chapter will elucidate the elements presented in the figure, articulating their interconnections and the resulting implications for grid management. By analyzing each component from resource availability to demand management and decision-making, this chapter will offer insights into the creation of a more resilient, efficient, and sustainable energy future.

M. H. Ahadi (✉)
Research and Education Promotion Association (REPA), Etobicoke, ON, Canada
e-mail: mahadi@repa.jp

M. S. S. Danish (ed.), *Unified Vision for a Sustainable Future*,
https://doi.org/10.1007/978-3-031-53574-1_4

2 Proposed Framework

The proposed framework shown in Fig. 1 [2, 6–13] presents a multifaceted approach designed to revolutionize the management of energy systems. By weaving together advanced machine learning algorithms with key economic principles, this framework aims not only to boost the efficiency and sustainability of the grid but also to fortify its reliability against fluctuating demands and unpredictable events [14, 15]. It intricately lays out strategies for optimizing supply chains, predictive maintenance, and demand-side management, ensuring that every facet of energy provision is enhanced through intelligent data analysis and real-time decision-making [16,

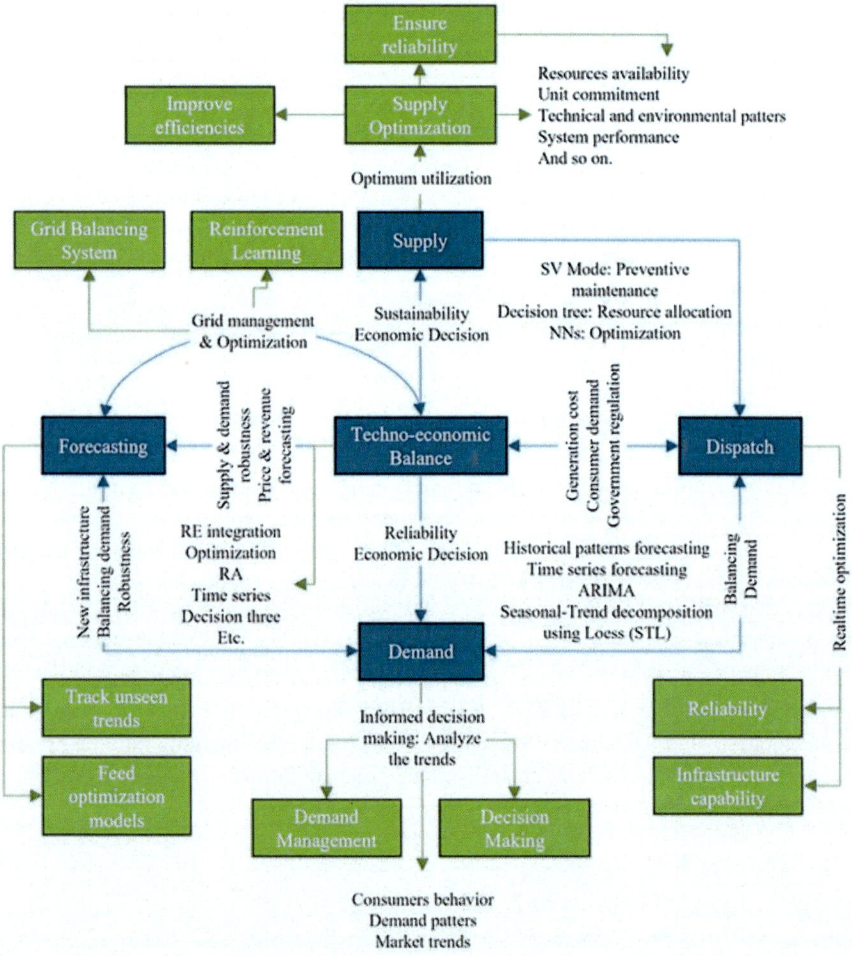

Fig. 1 Enhancing energy system efficiency through machine learning: a comprehensive framework for demand forecasting, resource allocation, and grid reliability

17]. This progressive framework stands as a blueprint for energy providers seeking to harness the transformative power of artificial intelligence, ultimately leading to a smarter, more resilient, and economically sound energy infrastructure. The components of this framework are briefly outlined in this section, and their in-depth details are discussed in the upcoming sections.

The framework depicted in Fig. 1 is a complex schematic that integrates machine learning (ML) into energy systems, focusing on enhancing grid efficiency and reliability through a techno-economic approach. Here is a detailed explanation of its components [18–25]:

– Grid Efficiency and Reliability
  - Improve efficiencies: It likely aimed at optimizing the energy grid to reduce waste and maximize the performance of the system.
  - Ensure reliability: Reliability is crucial for energy systems to provide a consistent and stable supply of power.
  - Supply optimization: Involves adjusting the supply side to meet demand efficiently.
  - Optimum utilization: Ensures that all resources are used in the best possible way.

– Components of the Framework
  - Grid balancing system: Ensures that supply and demand are in balance, which is critical for preventing power outages and maintaining a stable grid.
  - Forecasting: Predictive models such as time-series analysis and machine learning algorithms predict future demand and supply conditions.
  - Reinforcement learning: An area of ML focused on making a sequence of decisions. It can be used to optimize grid management by learning the best actions to take in real time.
  - Techno-economic balance: Balancing the technical aspects of the energy system with economic factors to make decisions that are both efficient and cost-effective.
  - Demand management: The process of controlling and influencing consumer demand to optimize energy consumption.
  - Decision-making: Analyzing trends and data to make informed choices about energy distribution and resource allocation.

– Machine Learning and Decision Support
  - Supply: Refers to the generation and provision of energy.
  - Sustainability economic decision: Decisions that take into account both environmental sustainability and economic viability.
  - Reliability economic decision: Economic decisions that focus on ensuring the reliability of the energy supply.
  - Infrastructure capability: The physical and technical capabilities of the energy infrastructure to handle supply and demand.

  - Feed optimization models: Using data and machine learning to continually improve optimization algorithms.
  - Track unseen trends: Detecting patterns and changes in energy consumption or supply that have not been previously identified.

- Forecasting and Analysis

  - RE integration optimization: Incorporating renewable energy sources into the grid in an optimized manner.
  - RA (Resource Allocation?): Allocating resources effectively, which is likely a part of optimization algorithms.
  - AutoRegressive Integrated Moving Average, or ARIMA, seasonal-trend decomposition using LOESS (STL): Statistical models used for forecasting. ARIMA is used for time-series forecasting, while STL is a method for decomposing a time series into seasonal, trend, and residual components.

- Maintenance and Resource Allocation

  - SV mode, preventive maintenance: A system mode focused on maintaining equipment before failures occur to prevent outages.
  - Decision tree, resource allocation: A decision support tool that uses a tree-like graph or model of decisions and their possible consequences to allocate resources efficiently.
  - Neural networks (NNs) optimization: Neural networks used for various optimization problems within the grid.

- Cost and Dispatch

  - Generation cost: The expenses associated with producing energy.
  - Consumer demand: The amount and pattern of energy usage by consumers.
  - Dispatch: The process of controlling the flow of electricity from power plants to consumers, ensuring that energy generation meets the demand.
  - Demand balancing: Adjusting the supply of energy to meet demand levels.
  - Real-time optimization: Using real-time data to continually optimize grid operations.

The proposed framework shown in Fig. 1 outlines a comprehensive approach to integrating ML techniques into energy system operations to enhance efficiency and reliability. It emphasizes the balance between technical capabilities and economic considerations, with a strong focus on predictive analytics and decision-making supported by data-driven insights.

3 Energy Efficiency

Efficiency in this context is not merely about conserving energy but is about optimizing the entire energy value chain from generation to consumption [26]. The introduction of smart technologies and analytics has paved the way for advanced metering, demand response programs, and grid automation, which contribute significantly to the operational efficiency of EMS [27].

Efficiency in grid management is tantamount to achieving a sustainable energy future [18]. It encompasses a myriad of strategies and technologies designed to ensure that energy production meets consumption demands with minimal waste and maximal cost-effectiveness [21]. Central to this endeavor are grid balancing systems, forecasting techniques, and the incorporation of renewable energy, all underpinned by sophisticated optimization algorithms and comprehensive regulatory assessment (RA) [9, 28]. Grid balancing systems are essential for the maintenance of a stable and reliable power supply [29]. They manage the constant and dynamic balance between energy supply and demand. These systems are increasingly complex and must accommodate a wide range of variables, ranging from fluctuating demand patterns to intermittent supply from renewable energy sources [15]. Advanced grid balancing encompasses real-time monitoring, rapid response capabilities, and predictive analyses to prevent imbalances that could lead to power outages or wasteful overproduction [30].

Forecasting plays a crucial role in grid efficiency [31]. By predicting demand and supply conditions, grid operators can make informed decisions to adjust energy flows, anticipate maintenance needs, and manage energy reserves. Techniques such as load forecasting, price forecasting, and generation forecasting enable the grid to operate more efficiently, thereby reducing costs and environmental impacts [32]. The integration of renewable energy sources, such as wind and solar power, introduces variability and unpredictability into the grid [17]. Efficient management of these sources requires innovative approaches to storage, like battery systems, and demand response programs that adjust usage during peak production times [33]. Effective integration of renewables is not just about adding green energy to the mix, but it is about rethinking the grid to be more flexible, adaptive, and smart [34]. The integration of such renewable energy sources, along with the advent of smart grid technologies, has introduced new challenges in grid management [35]. Cybersecurity, aging infrastructure, and the need for new regulatory frameworks are just some of the hurdles that must be overcome [36, 37]. In response, the energy sector is witnessing the rise of cutting-edge solutions, including AI-driven forecasting, grid-scale battery storage, and the Internet of Things (IoT) for enhanced monitoring and control [10, 38].

Optimization algorithms are the mathematical engines driving grid efficiency [12, 39]. They process vast amounts of data to determine the best course of action for energy distribution and generation. Coupled with RA, which ensures compliance with regulatory standards and environmental guidelines, these algorithms seek to find the optimal balance between efficiency, reliability, and sustainability [40]. Time-series analysis and decision trees are valuable tools for understanding and predicting grid behavior [41]. Time-series analysis helps in recognizing patterns and trends in energy usage over time, facilitating better planning and forecasting [42]. Decision trees, on the other hand, assist in making strategic decisions by mapping out the consequences of various actions, providing a structured approach to complex problem solving.

Unseen trends, such as unexpected changes in consumer behavior or sudden shifts in energy supply, can have significant impacts on grid efficiency [43, 44]. Identifying and understanding these trends allows for preemptive adjustments to grid operations. This foresight is critical to maintaining a resilient and efficient energy system, especially in the face of unforeseen events or shifts in the market [45]. The efficacy of optimization models hinges on the quality and timeliness of the data they receive. The advent of the Internet of Things (IoT) and smart sensors has revolutionized the collection and utilization of real-time data [46]. These technologies provide a continuous stream of information on grid performance, energy consumption, and other critical metrics. By leveraging this data, optimization models can perform with greater accuracy, adapting to changes as they happen and predicting future states of the grid with enhanced precision.

4 System Reliability

Reliability in power supply is a cornerstone of modern society, directly impacting everything from the smallest household routines to the largest industrial processes [47]. Ensuring continuous and reliable power supply requires a multifaceted approach, blending traditional strategies with innovative technologies and economic considerations [48]. Reliability, however, remains a nonnegotiable aspect of grid management [49]. The consequences of unreliability, ranging from blackouts to economic losses and safety hazards, underscore the importance of developing robust systems capable of withstanding both planned and unplanned stressors [50, 51]. This is further complicated by the variability introduced by renewable energy sources, necessitating innovative solutions for storage, demand forecasting, and responsive grid management [52, 53]. The core strategy to ensure reliability is the optimization of power supply [54]. This involves a suite of methods that range from predictive maintenance of power plants to the dynamic balancing of power loads across the grid. By employing advanced analytics, grid operators can preemptively identify potential disruptions and reroute power flows to mitigate the impact [55, 56]. Additionally, management strategies, such as incentivizing off-peak usage in residential and commercial sectors, play a crucial role in flattening demand spikes and ensuring a balanced load on the system [4, 57].

Reinforcement learning, a type of machine learning, is increasingly being utilized in grid management. This approach enables systems to learn and improve from experience without being explicitly programed [58]. In the context of grid management, reinforcement learning algorithms can optimize energy distribution in real time, continuously learning from grid behavior to make more accurate and robust decisions. This can lead to more efficient utilization of generation assets, reduced operational costs, and improved reliability, even under uncertain conditions [59]. Reliability is not only a technical challenge but also an economic challenge [60]. Achieving a techno-economic balance means ensuring that the technical solutions implemented for reliability also make economic sense [61]. Careful analysis and planning are required to balance the costs of new technologies and infrastructure with the benefits they provide in terms of reliability and efficiency. Investments in smart grid technologies, for example, are weighed against the expected improvements in grid performance and the economic benefits of avoiding outages and improving service quality [62].

Sustainability is an integral part of ensuring long-term reliability [17, 63]. This encompasses not only the environmental aspect of using renewable energy sources and reducing greenhouse gas emissions but also the sustainable management of grid resources [64]. Sustainable practices ensure that the grid is reliable not only for today's generation but also for future generations, which involves considering the life cycles of grid components and the sources of energy supplied.

Economic decisions in grid management involve a delicate balance between cost, reliability, and performance [65]. Decision-makers must consider the financial implications of investing in new infrastructure versus the costs associated with potential outages or inefficiencies. This includes the development of financial models that take into account the value of reliability to consumers and the economy at large, guiding investments in infrastructure upgrades, maintenance, and the integration of innovative technologies [66]. The capabilities of the existing infrastructure are a critical determinant of reliability. Aging infrastructure can be prone to failures, leading to outages and inefficiencies [67]. Upgrading this infrastructure is essential for improving reliability; however, it requires significant investment [49, 51]. New infrastructure is often designed with redundancy, fault tolerance, and the ability to integrate with renewable energy sources [68], which are vital for a reliable power supply in the context of increasing demand and the shift toward decarbonization [69].

5 Resource Allocation

In the domain of grid management, the assurance of resource availability and the strategic allocation of these resources is paramount to maintaining a stable and efficient power system [70, 71]. Resource assessment is the first critical step in grid management, involving a comprehensive evaluation of all available energy sources, including conventional fuels, renewable energy, and reserve capacities [72]. This assessment must not only take into account the current stock but also predict future stock availability by considering factors such as fuel supply contracts, renewable resource intermittency, and infrastructure constraints. The objective is to gain a complete picture of what resources are reliably available for power generation at any given time.

Unit commitment refers to the process of determining which power generation units should be turned on or off as well as their levels of production [73]. This decision-making process is complex and must consider the operational costs, start-up times, and the ramp-up capabilities of different units. Strategies for unit commitment also need to align with market conditions, demand forecasts, and regulatory requirements to ensure that the committed units can deliver the required power in the most cost-effective manner while maintaining grid stability [25].

Technical patterns in energy consumption, such as peak demand times and seasonal variations, must be aligned with environmental patterns that affect resource availability, such as solar irradiance and wind patterns for renewable energy generation [74]. Additionally, the environmental impact of resource extraction and energy production must be considered, ensuring compliance with environmental regulations and sustainability goals. This alignment is crucial for optimizing resource allocation and for the planning of infrastructure investments, such as transmission lines and energy storage systems [24, 75]. The ultimate goal of resource availability and allocation is to maintain optimal system performance. This involves ensuring that the power grid operates within its technical specifications, including frequency and voltage regulations [54, 76]. It also requires a proactive approach to the maintenance and management of the aging infrastructure to prevent failures [77]. Maintaining system performance also includes the flexibility to adapt to new technologies and the capacity to integrate distributed energy resources, such as rooftop solar and energy storage, which are increasingly important in modern grid management.

6 Maintenance and Optimization

Preventive maintenance and optimization are critical aspects of managing complex systems, such as electrical grids or industrial machinery [78]. SV (set value) mode, which likely refers to a specific mode in a system or a tool, is used for preventive maintenance [29]. In this context, SV mode might be a diagnostic or monitoring mode that helps in identifying potential issues before they become major problems. By analyzing data in this mode, maintenance teams can predict wear and tear, plan maintenance schedules, and reduce downtime. Decision trees are a type of machine learning algorithm that can be used for making decisions based on data [79]. In the context of resource allocation, decision trees can analyze various factors such as resource availability, demand, maintenance schedules, and operational efficiency. This helps in making informed decisions about where and how to allocate resources most effectively, ensuring optimal use of resources while minimizing waste and inefficiency [80]. Neural networks, a form of artificial intelligence, can be highly effective in optimizing grid management [81, 82]. They can process large amounts of data, learn from patterns, and make predictions or decisions based on that learning. In grid management, neural networks can be used for load forecasting, predicting equipment failure, optimizing energy distribution, and balancing supply and demand [14]. This helps in making the grid more efficient, reliable, and capable of handling various operational challenges.

7 Dispatch and Load Management

Dispatch and load management in power systems is a complex process that requires sophisticated analytical techniques and tools to ensure efficient, reliable, and sustainable electricity delivery [83]. The integration of advanced statistical methods with machine learning algorithms is becoming increasingly important in this field, enabling more accurate predictions and better decision-making [84]. Dispatch and load management involve balancing electricity supply with consumer demand. Careful planning and real-time adjustments are required to ensure efficient, reliable, and cost-effective power distribution. Generation costs play a crucial role in determining which power plants are dispatched at any given time. Typically, plants with lower operational costs are preferred. The analysis involves considering fuel costs, operational efficiency, and maintenance costs of different plants. Higher generation costs can lead to higher electricity prices and influence dispatch decisions, prioritizing more cost-effective sources [85].

Load management must balance the electricity demand of consumers with government regulations [86]. These regulations might pertain to environmental standards, energy efficiency, and renewable energy quotas. Utilities must navigate these requirements while ensuring that they meet consumer demand and maintain grid stability [15]. Accurate demand forecasting is essential for efficient load management. Techniques used include statistical methods, machine learning algorithms, and predictive analytics. Forecasting helps in anticipating future demand, planning generation, and avoiding energy shortages or wasteful excesses [42]. Analyzing historical data helps in understanding past consumption patterns, which is vital for improving forecasting accuracy [87]. Utilities examine data trends over time, considering factors like economic growth, population changes, and technological advancements that influence electricity usage.

Time-series forecasting involves analyzing data points collected at regular intervals to predict future values. ARIMA (AutoRegressive Integrated Moving Average) is a widely used statistical method in time-series forecasting [88]. It helps in understanding and forecasting future trends based on historical time-series data. Seasonal-trend decomposition using LOESS (STL) is a method used to decompose a time series into seasonal, trend, and residual components [89]. This technique is particularly useful in analyzing data with seasonal patterns, like electricity consumption, which often varies significantly with seasons due to heating in winter and cooling in summer.

8 Demand Management and Decision-Making

Demand management and decision-making are about understanding and anticipating needs and then strategically planning to meet those needs in the most efficient and effective manner [89]. This involves a combination of data analysis, market insight, and understanding of consumer behavior. Demand management in various sectors, particularly in energy and retail, is pivotal for balancing supply with consumer needs. Effective decision-making is central to this process.

Informed decision-making involves using data and analytics to understand and predict demand. This can include analyzing historical consumption data, market trends, and consumer behavior. Informed decisions help in optimizing resource allocation, reducing waste, and improving customer satisfaction. Understanding consumer behavior is key to effective demand management. This includes analyzing how consumers use products or services, their preferences, and how external factors like seasons, economic conditions, and social trends affect their behavior. By identifying patterns in consumer behavior, businesses and utilities can tailor their strategies to meet demand more efficiently. Market trend analysis involves studying changes in the market that could affect demand. This includes monitoring economic indicators, competitor activities, technological advancements, and changes in consumer preferences. Accurate evaluation of these trends enables prediction of future demand, allowing for proactive adjustments in supply chain, production, and distribution strategies.

9 Model Development and Validation in Energy Systems: A Data-Driven Approach

In the intricate arena of energy systems, the development of robust and reliable models is paramount [61]. The figure presented serves as a comprehensive roadmap, delineating a structured process for the creation, refinement, and validation of models that are fundamental to the management and optimization of energy systems [90]. This schematic elucidates a layered approach, commencing with the conceptualization of an energy system model and culminating in the rigorous selection and validation of a machine learning model [91]. The process begins by grappling with the inherent complexity of the energy system model, dissecting it into sub-models and identifying their dynamic characteristics [4]. These complex models are then subjected to a meticulous parameterization process, aligning with mathematical, physical, and control models. This level of abstraction is critical for encapsulating the multifaceted nature of energy systems in a theoretical framework, providing a blueprint for subsequent data-driven analysis [92].

Advancing into the data-driven model level encounters a cascade of data processing steps that are pivotal in shaping the eventual model. This includes the aggregation, optimization, and refinement of data, which is then funneled into machine learning datasets specifically curated for model training [10]. The interplay between model development and data processing is emphasized, highlighting the iterative nature of model tuning and the importance of data integrity.

The selection of a machine learning model is not merely a matter of algorithmic preference but is guided by preference sensitivity and model assessment requirements, ensuring that the chosen model aligns with the specificities and nuances of the energy system in question. The final stages of the process, validate, verify, and access, serve as the gatekeepers of model credibility, ensuring the model's operational efficacy within the real-world context of the energy system. Figure 2 illustrates a holistic and iterative approach to model development, integrating traditional modeling techniques with the latest in machine learning. It underscores the symbiotic relationship between data and models, where each iteration of the process enhances precision and reliability, driving toward an optimal solution that not only reflects the complex reality of energy systems but is also adaptable to their dynamic nature.

The framework shown in Fig. 2 emphasizes a systematic approach to developing a robust model for energy systems by leveraging data-driven techniques alongside traditional modeling methods. It ensures that the models developed are both theoretically sound and empirically validated, capable of handling the complexities and dynamics of modern energy systems that can be discussed briefly as follows:

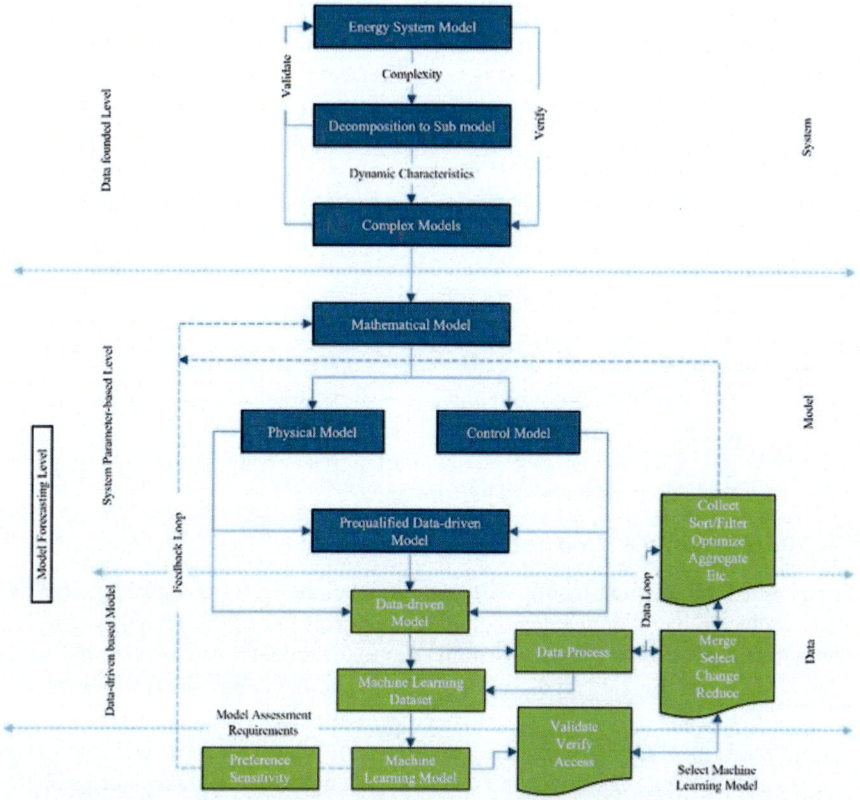

Fig. 2 An overview of the data-driven modeling and validation process through an exhaustive road map of machine learning applications in energy systems

- Model Development and Validation Process
  - Energy system model: This is the starting point of the framework. It represents the entire energy system with all its complexity.
  - Complexity: Recognizes that the energy system model is inherently complex.
  - Decomposition to submodel: This step involves breaking down the complex energy system model into more manageable submodels.
  - Dynamic characteristics: Focuses on the dynamic aspects of the energy system, which are likely time-variant.
  - Complex models: These are the detailed models that result from considering the energy system's complexity and dynamics.
- System Parameterization Level
  - Mathematical model: A theoretical representation of the energy system using mathematical equations and formulas.
  - Physical model: A representation that accounts for the physical laws and principles governing the energy system.
  - Control model: A model that includes control systems which could be used to manage the flow and distribution of energy.
- Data-Driven Model Level
  - Prequalified data-driven model: This refers to a model that has been developed based on historical data and has passed certain prequalification criteria to be considered valid.
  - Data-driven model: A model developed from data analytics, possibly including statistical models, machine learning algorithms, or other data processing techniques.
  - Machine learning dataset: The dataset specifically prepared and used to train machine learning models.
  - Data process.
  - Collect, sort/filter, optimize, and aggregate: These actions are part of the data processing workflow to prepare data for modeling.
  - Merge, select, change, and reduce: Further data manipulation tasks that refine the dataset for better model performance.
- Model Selection and Validation
  - Machine learning model: The specific machine learning algorithm or model selected based on the dataset and the requirements of the energy system.
  - Preference sensitivity: This could refer to the model's sensitivity to the preferences set by the modelers, such as what outcomes are most valued or prioritized.
  - Select machine learning model: The process of choosing the most appropriate machine learning model based on the system's needs and the data available.

  - Validate, verify, and access: Steps to ensure the model's validity and accuracy in representing the energy system and its ability to be accessed or utilized for the intended purpose.
- Feedback Loop
  - A feedback loop is shown from the data-driven model level back to the system parameterization level, indicating that the outcomes of the data-driven models can inform and refine the mathematical, physical, and control models.
- Overall Flow
  - Data input: This shows that data is an input to the entire process, informing each step of the model development.
  - Model: The final output is a validated model that can be used for the system's purposes.
  - System: This likely represents the actual energy system that the model aims to simulate or control.
  - Data output: The result of the data processing steps that feed into the model development process.

10 Discussions and Results

This study is embarked on an investigative journey to examine the integration of machine learning with energy systems, focusing on enhancing grid efficiency and reliability. The proposed framework shown in Fig. 1 has been dissected to reveal a meticulous strategy that aligns advanced machine learning algorithms with core economic principles, aiming to revolutionize energy system management. Throughout the energy efficiency exploration, we observed that the smart technology application, including advanced metering and grid automation, has significantly contributed to operational efficiencies. The grid balancing systems, underscored by forecasting and renewable energy incorporation, exhibited a potent combination when fortified by optimization algorithms and regulatory assessment. The role of forecasting emerged as a bedrock of grid efficiency, where machine learning algorithms predicted demand and supply conditions with commendable accuracy. This was particularly evident in the real-time optimization of grid operations, suggesting that forecasting techniques could substantially reduce costs and environmental impacts. The application of reinforcement learning algorithms demonstrated an unprecedented potential in real-time energy distribution optimization. The economic analysis of reliability interventions indicated a need for a delicate balance between technological investment and economic viability.

The discussions highlighted the critical nature of unseen trends, such as shifts in consumer behavior or sudden changes in energy supply, which have profound implications for grid efficiency. The Internet of Things (IoT) and smart sensors have revolutionized data collection, enabling optimization models to adapt and predict grid states with enhanced precision. The results indicate that the integration of machine learning with energy system models yields significant improvements in efficiency and reliability. The iterative nature of model refinement, driven by a symbiotic relationship between data and models, has led to an optimal solution that reflects the complex realities of energy systems and their dynamic nature.

11 Conclusion

This study presents an exhaustive exploration into the integration of machine learning and energy system management, proposing a framework that promises enhanced grid efficiency and reliability. The discussions confirm the transformative potential of machine learning, underpinned by the need for theoretical soundness and empirical validation of the developed models.

The study addresses the necessity of meticulous data preprocessing, model selection tailored to system complexity, and continuous refinement processes. Looking ahead, there is an avenue for further exploration of applying this framework across a broader spectrum of machine learning techniques and energy system scenarios. The groundwork laid by this study serves as a stepping stone for future research, guiding energy providers toward a new era marked by intelligence, resilience, and economic soundness in energy management. The integration of machine learning is not just an advancement in technology but it is a leap toward a sustainable energy future.

References

1. Ackermann, T., Andersson, G., Söder, L.: Distributed generation: a definition. Electr. Power Syst. Res. **57**, 195–204 (2001). https://doi.org/10.1016/S0378-7796(01)00101-8
2. Danish, M.S.S., Nazari, Z., Senjyu, T.: AI-coherent data-driven forecasting model for a combined cycle power plant. Energy Convers. Manag. **286**, 117063 (2023). https://doi.org/10.1016/j.enconman.2023.117063
3. Al-Najideen, M.I., Alrwashdeh, S.S.: Design of a solar photovoltaic system to cover the electricity demand for the faculty of engineering- Mu'tah University in Jordan. Resour.-Effic. Technol. **3**, 440–445 (2017). https://doi.org/10.1016/j.reffit.2017.04.005
4. Furukakoi, M., Sediqi, M.M., Senjyu, T., Danish, M.S.S., Howlader, A.M., Hassan, M.A.M., Funabashi, T.: Optimum capacity of energy storage system considering solar radiation forecast error and demand response. In: 2017 IEEE 3rd International Future Energy Electronics Conference and ECCE Asia (IFEEC 2017 – ECCE Asia), pp. 997–1001. IEEE, Kaohsiung (2017). https://doi.org/10.1109/IFEEC.2017.7992177
5. Danish, M.S.S., Senjyu, T.S.: Green building efficiency and sustainability indicators. In: Green Building Management and Smart Automation, pp. 128–145. IGI Global, Hershey (2020). https://doi.org/10.4018/978-1-5225-9754-4.ch006
6. Danish, M.S.S., Zaheb, H., Sabory, N.R., Tomonobu, S., Ahmadi, M., Sadat, S.H.: Empowering Developing Nations and Sustainable Development: Case Studies and Synthesis. REPA – Research and Education Promotion Association, Japan (2020)

7. Danish, M.S.S., Senjyu, T.S.: System of green resilience eco-oriented land uses in urban socio-ecosystems. In: Eco-Friendly Energy Processes and Technologies for Achieving Sustainable Development, pp. 1–23. IGI Global, Hershey (2021). https://doi.org/10.4018/978-1-7998-4915-5
8. Ahmadi, M., Adewuyi, O.B., Danish, M.S.S., Mandal, P., Yona, A., Senjyu, T.: Optimum coordination of centralized and distributed renewable power generation incorporating battery storage system into the electric distribution network. Int. J. Electr. Power Energy Syst. **125**, 106458 (2021). https://doi.org/10.1016/j.ijepes.2020.106458
9. Danish, M.S.S.: A framework for modeling and optimization of data-driven energy systems using machine learning. IEEE Trans. Artif. Intell. **1–10** (2023). https://doi.org/10.1109/TAI.2023.3322395
10. Danish, M.S.S.: AI and expert insights for sustainable energy future. Energies. **16**, 3309 (2023). https://doi.org/10.3390/en16083309
11. Danish, M.S.S., Bhattacharya, A., Stepanova, D., Mikhaylov, A., Grilli, M.L., Khosravy, M., Senjyu, T.: A systematic review of metal oxide applications for energy and environmental sustainability. Metals. **10**, 1604 (2020). https://doi.org/10.3390/met10121604
12. Danish, M.S.S., Senjyu, T.: AI-enabled energy policy for a sustainable future. Sustain. For. **15**, 7643 (2023). https://doi.org/10.3390/su15097643
13. Danish, M.S.S., Elsayed, M.E.L., Ahmadi, M., Senjyu, T., Karimy, H., Zaheb, H.: A strategic-integrated approach for sustainable energy deployment. Energy Rep. **6**, 40–44 (2020). https://doi.org/10.1016/j.egyr.2019.11.039
14. Danish, M.S.S., Senjyu, T.: Shaping the future of sustainable energy through AI-enabled circular economy policies. Circ. Econ. **2**, 100040 (2023). https://doi.org/10.1016/j.cec.2023.100040
15. Danish, M.S.S., Senjyu, T., Danish, S.M.S., Sabory, N.R., Narayanan, K., Mandal, P.: A recap of voltage stability indices in the past three decades. Energies. **12**, 1544 (2019). https://doi.org/10.3390/en12081544
16. Danish, M.S.S., Senjyu, T., Faisal, N., Stannikzai, M.Z., Nazari, A.M., Vargas-Hernández, J.G.: A review on environmental-friendly energy multidisciplinary exposition from goals to action. J. Environ. Sci. Revolut. **2**, 1–9 (2021). https://doi.org/10.37357/1068/jesr.2.1.01
17. Danish, M.S.S., Senjyu, T., Funabashia, T., Ahmadi, M., Ibrahimi, A.M., Ohta, R., Rashid Howlader, H.O., Zaheb, H., Sabory, N.R., Sediqi, M.M.: A sustainable microgrid: a sustainability and management-oriented approach. Energy Procedia. **159**, 160–167 (2019). https://doi.org/10.1016/j.egypro.2018.12.045
18. Danish, M.S.S., Senjyu, T., Ibrahimi, A.M., Ahmadi, M., Howlader, A.M.: A managed framework for energy-efficient building. J. Build. Eng. **21**, 120–128 (2019). https://doi.org/10.1016/j.jobe.2018.10.013
19. Danish, M.S.S., Senjyu, T., Ibrahimi, A.M., Bhattacharya, A., Nazari, Z., Danish, S.M.S., Ahmadi, M.: Sustaining energy systems using metal oxide composites as photocatalysts. J. Sustain. Energy Revolut. **2**(6–15), 6 (2021). https://doi.org/10.37357/1068/jser.2.1.02
20. Danish, M.S.S., Senjyu, T., Nazari, M., Zaheb, H., Nassor, T.S., Danish, S.M.S., Karimy, H.: Smart and sustainable building appraisal. J. Sustain. Energy Revolut. **2**, 1–5 (2021). https://doi.org/10.37357/1068/jser.2.1.01
21. Danish, M.S.S., Senjyu, T., Zaheb, H., Sabory, N.R., Ibrahimi, A.M., Matayoshi, H.: A novel transdisciplinary paradigm for municipal solid waste to energy. J. Clean. Prod. **233**, 880–892 (2019). https://doi.org/10.1016/j.jclepro.2019.05.402
22. Danish, M.S.S., Yona, A., Senjyu, T.: Pre-design and life cycle cost analysis of a hybrid power system for rural and remote communities in Afghanistan. J. Eng. IET. **2014**, 438–444 (2014). https://doi.org/10.1049/joe.2014.0172
23. Furukakoi, M., Danish, M.S.S., Howlader, A.M., Senjyu, T.: Voltage stability improvement of transmission systems using a novel shunt capacitor control. Int. J. Emerg. Electr. Power Syst. **19**, 1–12 (2018). https://doi.org/10.1515/ijeeps-2017-0112
24. Ibrahimi, A.M., Howlader, H.O.R., Danish, M.S.S., Shigenobu, R., Sediqi, M.M., Senjyu, T.: Optimal unit commitment with concentrated solar power and thermal energy storage in Afghanistan electrical system. Int. J. Emerg. Electr. Power Syst. **20** (2019). https://doi.org/10.1515/ijeeps-2018-0264

25. Ibrahimi, A.M., Sediqi, M.M., Howlader, H.O.R., Danish, M.S.S., Chakraborty, S., Senjyu, T.: Generation expansion planning considering renewable energy integration and optimal unit commitment: a case study of Afghanistan. AIMS Energy. **7**, 441–464 (2019). https://doi.org/10.3934/energy.2019.4.441
26. Danish, M.S.S., Senjyu, T., Ahmadi, M., Ludin, G.A., Ahadi, M.H., Karimy, H., Khosravy, M.: A review on energy efficiency for pathetic environmental trends mitigation. J. Sustain. Outreach. **2**, 1–8 (2021). https://doi.org/10.37357/1068/jso.2.1.01
27. Kräuchi, P., Dahinden, C., Jurt, D., Wouters, V., Menti, U.-P., Steiger, O.: Electricity consumption of building automation. Energy Procedia. **122**, 295–300 (2017). https://doi.org/10.1016/j.egypro.2017.07.325
28. Karagiannopoulos, S., Dobbe, R., Aristidou, P., Callaway, D., Hug, G.: Data-driven control design schemes in active distribution grids: capabilities and challenges. In: 2019 IEEE Milan PowerTech, pp. 1–6 (2019). https://doi.org/10.1109/PTC.2019.8810586
29. Hernández-Callejo, L., Gallardo-Saavedra, S., Alonso-Gómez, V.: A review of photovoltaic systems: design, operation and maintenance. Sol. Energy. **188**, 426–440 (2019). https://doi.org/10.1016/j.solener.2019.06.017
30. Furukakoi, M., Adewuyi, O.B., Danish, M.S.S., Howlader, A.M., Senjyu, T., Funabashi, T.: Critical boundary index (CBI) based on active and reactive power deviations. Int. J. Electr. Power Energy Syst. **100**, 50–57 (2018). https://doi.org/10.1016/j.ijepes.2018.02.010
31. Ahmed, R., Sreeram, V., Mishra, Y., Arif, M.D.: A review and evaluation of the state-of-the-art in PV solar power forecasting: techniques and optimization. Renew. Sust. Energ. Rev. **124**, 109792 (2020). https://doi.org/10.1016/j.rser.2020.109792
32. Shams, S., Danish, M.S.S., Sabory, N.R.: Solar energy market and policy instrument analysis to support sustainable development. In: Danish, M.S.S., Senjyu, T., Sabory, N.R. (eds.) Sustainability Outreach in Developing Countries, pp. 113–132. Springer, Singapore (2021). https://doi.org/10.1007/978-981-15-7179-4_8
33. Sward, J.A., Siff, J., Gu, J., Zhang, K.M.: Strategic planning for utility-scale solar photovoltaic development – historical peak events revisited. Appl. Energy. **250**, 1292–1301 (2019). https://doi.org/10.1016/j.apenergy.2019.04.178
34. Nazim, S.F., Danish, M.S.S., Senjyu, T.: A brief review of the future of smart mobility using 5G and IoT. J. Sustain. Outreach. **19–30**, 10.37357/1068/jso/3.1.02 (2022)
35. Brenna, M., Falvo, M.C., Foiadelli, F., Martirano, L., Poli, D.: Sustainable energy microsystem (SEM): preliminary energy analysis. In: 2012 IEEE PES Innovative Smart Grid Technologies (ISGT), pp. 1–6. IEEE, Washington (2012). https://doi.org/10.1109/ISGT.2012.6175735
36. Berizzi, A., Finazzi, P.: First and second order methods for voltage collapse assessment and security enhancement. IEEE Trans. Power Syst. **13**, 543–551 (1998). https://doi.org/10.1109/59.667380
37. Gao, B., Morison, G.K., Kundur, P.: Voltage stability evaluation using modal analysis. IEEE Trans. Power Syst. **7**, 1529–1542 (1992). https://doi.org/10.1109/59.207377
38. Danish, M.S.S.: AI in energy: overcoming unforeseen obstacles. AI. **4**, 406–425 (2023). https://doi.org/10.3390/ai4020022
39. Abdolrasol, M.G.M., Hussain, S.M.S., Ustun, T.S., Sarker, M.R., Hannan, M.A., Mohamed, R., Ali, J.A., Mekhilef, S., Milad, A.: Artificial neural networks based optimization techniques: a review. Electronics. **10**, 2689 (2021). https://doi.org/10.3390/electronics10212689
40. Danish, M.S.S., Sabory, N.R., Ibrahimi, A.M., Senjyu, T., Ahadi, M.H., Stanikzai, M.Z.: A concise overview of energy development within sustainability requirements. In: Danish, M.S.S., Senjyu, T., Sabory, N.R. (eds.) Sustainability Outreach in Developing Countries, pp. 15–27. Springer, Singapore (2021). https://doi.org/10.1007/978-981-15-7179-4_2
41. Alpaydin, E.: Introduction to Machine Learning. The MIT Press, Cambridge (2009)
42. Gutierrez-Corea, F.-V., Manso-Callejo, M.-A., Moreno-Regidor, M.-P., Manrique-Sancho, M.-T.: Forecasting short-term solar irradiance based on artificial neural networks and data from neighboring meteorological stations. Sol. Energy. **134**, 119–131 (2016). https://doi.org/10.1016/j.solener.2016.04.020

43. Sufizada, Z., Oryakheill, A.A., Kohnaward, M.H., Fazli, N., Zadran, H., Sabory, N.R., Danish, M.S.S.: From consumers to producers: energy efficiency as a tool for sustainable development in the context of informal settlements. In: Danish, M.S.S., Senjyu, T., Sabory, N.R. (eds.) Sustainability Outreach in Developing Countries, pp. 169–187. Springer, Singapore (2021). https://doi.org/10.1007/978-981-15-7179-4_11
44. Senjyu, T., Takara, H., Uezato, K., Funabashi, T.: One-hour-ahead load forecasting using neural network. IEEE Trans. Power Syst. **17**, 113–118 (2002). https://doi.org/10.1109/59.982201
45. Bhattacharyya, S.C.: Energy Economics: Concepts, Issues, Markets and Governance. Springer, London (2019)
46. Brahma, S., Kavasseri, R., Cao, H., Chaudhuri, N.R., Alexopoulos, T., Cui, Y.: Real-time identification of dynamic events in power systems using PMU data, and potential applications – models, promises, and challenges. IEEE Trans. Power Deliv. **32**, 294–301 (2017). https://doi.org/10.1109/TPWRD.2016.2590961
47. Khalid, H.M., Flitti, F., Mahmoud, M.S., Hamdan, M.M., Muyeen, S.M., Dong, Z.Y.: Wide area monitoring system operations in modern power grids: a median regression function-based state estimation approach towards cyber attacks. Sustain. Energy Grids Netw. **34**, 101009 (2023). https://doi.org/10.1016/j.segan.2023.101009
48. Danish, M.S.S., Senjyu, T. (eds.): Eco-Friendly and Agile Energy Strategies and Policy Development. IGI Global, Hershey (2022)
49. Danish, M.S.S.: Voltage Stability in Electric Power System: a Practical Introduction. Logos Verlag Berlin GmbH, Berlin (2015)
50. Blackouts, P.W.: A three-stage procedure for controlled islanding to prevent wide-area blackouts. Energies. **11**, 1–15 (2018). https://doi.org/10.3390/en11113066
51. Danish, M.S.S., Yona, A., Senjyu, T.: A review of voltage stability assessment techniques with an improved voltage stability indicator. Int. J. Emerg. Electr. Power Syst. **16**, 107–115 (2015). https://doi.org/10.1515/ijeeps-2014-0167
52. Danish, S.M.S., Ahmadi, M., Danish, M.S.S., Mandal, P., Yona, A., Senjyu, T.: A coherent strategy for peak load shaving using energy storage systems. J. Energy Storage. **32**, 101823 (2020). https://doi.org/10.1016/j.est.2020.101823
53. Ahmadi, M., Danish, M.S.S., Lotfy, M.E., Yona, A., Hong, Y.-Y., Senjyu, T.: Multi-objective time-variant optimum automatic and fixed type of capacitor bank allocation considering minimization of switching steps. AIMS Energy. **7**, 792 (2019). https://doi.org/10.3934/energy.2019.6.792
54. Danish, S.M.S., Shigenobu, R., Kinjo, M., Mandal, P., Krishna, N., Hemeida, A.M., Senjyu, T.: A real distribution network voltage regulation incorporating auto-tap-changer pole transformer multiobjective optimization. Appl. Sci. **9**, 2813 (2019). https://doi.org/10.3390/app9142813
55. Danish, M.S.S., Sabory, N.R., Funabashi, T., Danish, S.M.S., Noorzad, A.S., Yona, A., Senjyu, T.: Comparative analysis of load flow calculation methods with considering the voltage stability constraints. In: 2016 IEEE International Conference on Power and Energy (PECon), pp. 250–255. IEEE, Melaka (2016). https://doi.org/10.1109/PECON.2016.7951568
56. Yang, H., Wen, F., Wang, L.: Newton-Raphson on power flow algorithm and Broyden method in the distribution system. In: 2008 IEEE 2nd International Power and Energy Conference, pp. 1613–1618 (2008), https://doi.org/10.1109/PECON.2008.4762737
57. Sagara, M., Shigenobu, R., Adewuyi, O.B., Yona, A., Senjyu, T., Danish, M.S.S., Funabashi, T.: Voltage stability improvement by demand response. In: TENCON 2017–2017 IEEE Region 10 Conference, pp. 2144–2149. IEEE, Penang (2017). https://doi.org/10.1109/TENCON.2017.8228215
58. Glavic, M., Fonteneau, R., Ernst, D.: Reinforcement learning for electric power system decision and control: past considerations and perspectives. IFAC-Pap. **50**, 6918–6927 (2017). https://doi.org/10.1016/j.ifacol.2017.08.1217
59. Razmjoo, A.A., Sumper, A., Davarpanah, A.: Energy sustainability analysis based on SDGs for developing countries. Energy Sources Part Recovery Util. Environ. Eff. **42**, 1041–1056 (2020). https://doi.org/10.1080/15567036.2019.1602215

60. Agajie, T.F., Ali, A., Fopah-Lele, A., Amoussou, I., Khan, B., Velasco, C.L.R., Tanyi, E.: A comprehensive review on techno-economic analysis and optimal sizing of hybrid renewable energy sources with energy storage systems. Energies. **16**, 642 (2023). https://doi.org/10.3390/en16020642
61. Mirbarati, S.H., Heidari, N., Nikoofard, A., Danish, M.S.S., Khosravy, M.: Techno-economic-environmental energy Management of a Micro-Grid: a mixed-integer linear programming approach. Sustain. For. **14**, 15036 (2022). https://doi.org/10.3390/su142215036
62. Lu, R., Hong, S.H.: Incentive-based demand response for smart grid with reinforcement learning and deep neural network. Appl. Energy. **236**, 937–949 (2019). https://doi.org/10.1016/j.apenergy.2018.12.061
63. Danish, M.S.S.: Exploring metal oxides for hydrogen evolution reaction (HER) in the field of nanotechnology. RSC Sustain. **1**, 2180 (2023). https://doi.org/10.1039/D3SU00179B
64. Gorjian, S., Calise, F., Kant, K., Ahamed, M.S., Copertaro, B., Najafi, G., Zhang, X., Aghaei, M., Shamshiri, R.R.: A review on opportunities for implementation of solar energy technologies in agricultural greenhouses. J. Clean. Prod. **285**, 124807 (2021). https://doi.org/10.1016/j.jclepro.2020.124807
65. Waas, T., Hugé, J., Block, T., Wright, T., Benitez-Capistros, F., Verbruggen, A.: Sustainability assessment and indicators: tools in a decision-making strategy for sustainable development. Sustain. For. **6**, 5512–5534 (2014). https://doi.org/10.3390/su6095512
66. Adalı, Z., Danish, M.S.S.: Investigation of the nexus between the electricity consumption and the ecological footprint. In: Dinçer, H., Yüksel, S. (eds.) Circular Economy and the Energy Market: Achieving Sustainable Economic Development Through Energy Policy, pp. 79–89. Springer International Publishing, Cham (2022). https://doi.org/10.1007/978-3-031-13146-2_7
67. Ohno, T., Imai, S.: The 1987 Tokyo blackout. In: 2006 IEEE PES Power Systems Conference and Exposition, pp. 314–318. IEEE (2006). https://doi.org/10.1109/PSCE.2006.296325
68. Nimpitiwan, N., Heydt, G.T., Ayyanar, R., Suryanarayanan, S.: Fault current contribution from synchronous machine and inverter based distributed generators. IEEE Trans. Power Deliv. **22**, 634–641 (2007). https://doi.org/10.1109/TPWRD.2006.881440
69. Castrillón-Mendoza, R., Rey-Hernández, J.M., Rey-Martínez, F.J.: Industrial decarbonization by a new energy-baseline methodology. Case Study. Sustain. **12**, 1960 (2020). https://doi.org/10.3390/su12051960
70. Danish, M.S.S., Senjyu, T., Sabory, N.R., Danish, S.M.S., Ludin, G.A., Noorzad, A.S., Yona, A.: Afghanistan's aspirations for energy independence: water resources and hydropower energy. Renew. Energy. **113**, 1276–1287 (2017). https://doi.org/10.1016/j.renene.2017.06.090
71. Driesen, J., Katiraei, F.: Design for distributed energy resources. IEEE Power Energy Mag. **6**, 30–40 (2008). https://doi.org/10.1109/MPE.2008.918703
72. Kumar, R., Ojha, K., Ahmadi, M.H., Raj, R., Aliehyaei, M., Ahmadi, A., Nabipour, N.: A review status on alternative arrangements of power generation energy resources and reserve in India. Int. J. Low-Carbon Technol. **15**, 224–240 (2020). https://doi.org/10.1093/ijlct/ctz066
73. Duan, C., Jiang, L., Fang, W., Liu, J.: Data-driven Affinely adjustable distributionally robust unit commitment. IEEE Trans. Power Syst. **33**, 1385–1398 (2018). https://doi.org/10.1109/TPWRS.2017.2741506
74. Tong, W., Mu, D., Zhao, F., Mendis, G.P., Sutherland, J.W.: The impact of cap-and-trade mechanism and consumers' environmental preferences on a retailer-led supply chain. Resour. Conserv. Recycl. **142**, 88–100 (2019). https://doi.org/10.1016/j.resconrec.2018.11.005
75. Ahmadi, M., Lotfy, M.E., Howlader, A.M., Yona, A., Senjyu, T.: Centralised multi-objective integration of wind farm and battery energy storage system in real-distribution network considering environmental, technical and economic perspective. Transm. Distrib. IET Gener. **13**, 5207–5217 (2019). https://doi.org/10.1049/iet-gtd.2018.6749
76. El-Moursi, M.S., Sharaf, A.M.: Novel controllers for the 48-pulse VSC STATCOM and SSSC for voltage regulation and reactive power compensation. IEEE Trans. Power Syst. **20**, 1985–1997 (2005). https://doi.org/10.1109/TPWRS.2005.856996

77. Kurita, A., Sakuraj, T.: The power system failure on July 23, 1987 in Tokyo. In: Proceedings of the 27th IEEE Conference on Decision and Control, pp. 2093–2097. IEEE, Austin (1988). https://doi.org/10.1109/CDC.1988.194703
78. Aboagye, B., Gyamfi, S., Ofosu, E.A., Djordjevic, S.: Investigation into the impacts of design, installation, operation and maintenance issues on performance and degradation of installed solar photovoltaic (PV) systems. Energy Sustain. Dev. **66**, 165–176 (2022). https://doi.org/10.1016/j.esd.2021.12.003
79. Goodfellow, I., Bengio, Y., Courville, A.: Deep Learning: Adaptive Computation and Machine Learning Series. The MIT Press, Cambridge (2016)
80. Ahmadi, M., Lotfy, M.E., Shigenobu, R., Howlader, A.M., Senjyu, T.: Optimal sizing of multiple renewable energy resources and PV inverter reactive power control encompassing environmental, technical, and economic issues. IEEE Syst. J. **13**, 3026–3037 (2019). https://doi.org/10.1109/JSYST.2019.2918185
81. Fulginei, F.R., Salvini, A., Parodi, M.: Learning optimization of neural networks used for MIMO applications based on multivariate functions decomposition. Inverse Probl. Sci. Eng. **20**, 29–39 (2012). https://doi.org/10.1080/17415977.2011.629047
82. Runge, J., Zmeureanu, R.: Forecasting energy use in buildings using artificial neural networks: a review. Energies. **12**, 3254 (2019). https://doi.org/10.3390/en12173254
83. Kessel, P., Glavitsch, H.: Estimating the voltage stability of a power system. IEEE Trans. Power Deliv. **1**, 346–354 (1986). https://doi.org/10.1109/TPWRD.1986.4308013
84. Wang, J.-J., Jing, Y.-Y., Zhang, C.-F., Zhao, J.-H.: Review on multi-criteria decision analysis aid in sustainable energy decision-making. Renew. Sust. Energ. Rev. **13**, 2263–2278 (2009). https://doi.org/10.1016/j.rser.2009.06.021
85. Cheng, C., Liu, B., Chau, K.-W., Li, G., Liao, S.: China's small hydropower and its dispatching management. Renew. Sust. Energ. Rev. **42**, 43–55 (2015). https://doi.org/10.1016/j.rser.2014.09.044
86. Berizzi, A., Marannino, P., Merlo, M., Pozzi, M., Zanellini, F.: Steady-state and dynamic approaches for the evaluation of loadability margins in the presence of secondary voltage regulation. IEEE Trans. Power Syst. **19**, 1048–1057 (2004). https://doi.org/10.1109/TPWRS.2004.825869
87. Bourdeau, M., Zhai, X., qiang Nefzaoui, E., Guo, X., Chatellier, P.: Modeling and forecasting building energy consumption: a review of data-driven techniques. Sustain. Cities Soc. **48**, 101533 (2019). https://doi.org/10.1016/j.scs.2019.101533
88. Adli, H.K., Husin, K.A.K., Hanafiah, N.H.M., Remli, M.A., Ernawan, F., Wirawan, P.W.: Forecasting and analysis of solar power output from integrated solar energy and IoT system. In: 2021 5th International Conference on Informatics and Computational Sciences (ICICoS), pp. 222–226. IEEE, Semarang (2021). https://doi.org/10.1109/ICICoS53627.2021.9651831
89. Wu, D., Wang, Y., Li, L., Lu, P., Liu, S., Dai, C., Pan, Y., Zhang, Z., Lin, Z., Yang, L.: Demand response ability evaluation based on seasonal and trend decomposition using LOESS and S–G filtering algorithms. Energy Rep. **8**, 292–299 (2022). https://doi.org/10.1016/j.egyr.2022.02.139
90. Danish, M.S.S., Matayoshi, H., Howlader, H.O.R., Chakraborty, S., Mandal, P., Senjyu, T.: Microgrid planning and design: resilience to sustainability. In: 2019 IEEE PES GTD Grand International Conference and Exposition Asia (GTD Asia), pp. 253–258. IEEE, Bangkok (2019). https://doi.org/10.1109/GTDAsia.2019.8716010
91. Susowake, Y., Ibrahimi, A.M., Danish, M.S.S., Senjyu, T., Howlader, A.M., Mandal, P.: Multi-objective design of power system introducing seawater electrolysis plant for remote Island. In: 2018 IEEE Innovative Smart Grid Technologies – Asia (ISGT Asia), pp. 909–911. IEEE, Singapore (2018). https://doi.org/10.1109/ISGT-Asia.2018.8467912
92. Sabory, N.R., Senjyu, T., Danish, M.S.S., Ahmadi, M., Zaheb, H., Halim, M.: A framework for integration of smart and sustainable energy systems in urban planning processes of low-income developing countries: Afghanistan case. Sustain. For. **13**, 8428 (2021). https://doi.org/10.3390/su13158428

Renewable Energy and Power Flow in Microgrids: An Introductory Perspective

Mohammad Hamid Ahadi, Hameedullah Zaheb, and Tomonobu Senjyu

1 Introduction

The global population is estimated to increase to 8.6 billion by 2035. Undoubtedly, there will be a significant development in technology, economic growth, and energy consumption, in which the economic growth is correlative to the energy consumption rate [1]. Unlike previous non-energy resources, the main drivers for the utilization and exploitation of renewable energy resources are global energy security, mitigation of greenhouse gases, and the economic impact of renewable energy resources. The concept of clean energy as an alternative to nonrenewable energy resources can be interpreted as a renewable energy supply with less environmental impact (clean and green) and deployment of renewable energy technologies that are more efficient (cost-effective and efficient energy production). In recent years, a large amount of investment has been allocated to renewable energy supply technologies. This fact shows that renewable energy is increasingly being deployed.

Microgrids featured with diverse techno-economic perfections of system expansion and green energy integration flexibility with high efficiency, operation stability, local circular economy resiliency, and long-run sustainability in a dynamic nature. The global energy utility sector is rapidly transitioning toward automated and managed microgrids, marking a significant step toward the development of smart grids.

M. H. Ahadi (✉)
Research and Education Promotion Association (REPA), Etobicoke, ON, Canada
e-mail: mahadi@repa.jp

H. Zaheb
Department of Energy Engineering, Faculty of Engineering, Kabul University, Kabul, Afghanistan

T. Senjyu
Department of Electrical and Electronic Engineering, University of the Ryukyus, Okinawa, Japan

M. S. S. Danish (ed.), *Unified Vision for a Sustainable Future*,
https://doi.org/10.1007/978-3-031-53574-1_5

Microgrids are small-scale power systems featuring complex distribution configurations like interconnected, radial, and hybrid setups [2]. They integrate various elements such as generation sources (including photovoltaic (PV), wind, fuel cells, bioenergy, and combined heat and power), storage systems, diesel generators, loads, and advanced monitoring, control, and automation systems [3]. This setup aims to enhance the reliability of utility services to customers. Microgrids offer numerous advantages, such as improved resilience against disruptions, minimized system downtime, efficient emergency response, adaptability for expansion, seamless integration of renewable energy, and the ability to quickly incorporate customized and virtual automation features in buildings. The specific components of a microgrid vary based on business needs and operational behaviors.

Distributed generation (DG) systems are integral to microgrids, generating electricity close to the load [4]. This proximity to the load allows DGs to operate with high quality and stability while minimizing transmission losses [5].

Voltage source inverters play a crucial role in the operation of multiterminal AC-DC systems by converting and regulating power flow between the source, load, and the system [6]. These systems often incorporate different protection devices and equipment, e.g., hybrid circuit breakers, combining capacitive, reactive, inductive, and reactance elements to manage effective resistance [7]. Improving inverter efficiency and arbitrary switching frequencies from 1 to 50 kHz are examined in most studies to minimize output current ripple and ensure purity of sine wave output. However, semiconductor switching technologies demonstrate a variety of applications in which most insulated-gate bipolar transistor (IGBT) can be applied for a range of 20–50 kHz, with a preferred 30 kHz application [8]. Nearly 30% of conceptions utilize energy after it has undergone a power conversion process, indicating a significant transformation from its original generation form to a usable state for various applications [9]. This percentage denotes that power converters (converters and inverters) play a significant role in power systems.

Various approaches can be taken to optimize the system's performance in a hybrid AC-DC microgrid power flow. The unified approach combines AC and DC components into the same matrix, enabling simultaneous analysis [4]. On the other hand, the sequential approach solves AC-DC load flows separately, where DC load flow is injected into the AC load flow calculation [10]. The eliminated approach focuses on achieving convergence between the two systems for improved efficiency and stability [11].

Power flow (load flow/network flow) is the comprehensive analysis of electrical energy flow within an interconnected power system, accounting for various components and their interactions and integrating renewable energy sources, storage, and demand management to achieve stable and efficient operating conditions [12].

In the multifaceted realm of power system management, a broad array of components, operations, analytical techniques, and planning strategies are integral to the effective functioning of the electrical grid. This overview categorizes the essential items and activities involved in the seamless operation and robust management of power systems. The classification includes the physical components of power systems, such as generators and transmission lines, which form the backbone of electrical infrastructure. It also encompasses the dynamic aspects of system operation and control, which are vital for maintaining voltage and frequency stability. Analytical methods for network analysis and modeling are crucial for ensuring system reliability and efficiency, while planning and optimization are leveraged to predict and shape the future of power grid operations. Lastly, interconnected systems and market mechanisms are explored, recognizing the growing complexity and interdependence of modern power systems, which must adhere to a variety of standards and respond to economic signals. This structured categorization in Table 1 provides a comprehensive guide to the diverse elements that underpin contemporary power system management, from the tangible assets to the regulatory and economic environments in which they operate.

Table 1 Categorization of power system components

Category	Items
Power system components	Generators
	Loads
	Transmission lines
	Transformers
	Reactive power compensation devices
	Renewable energy sources
	Energy storage devices
	Distributed generation
	Protection devices
System operation and control	Voltage control
	Frequency control
	Demand response
	Power factor correction
	Power quality
	Monitoring and control systems
	Short-circuit analysis
Network analysis and modeling	Contingency analysis
	System topology
	System constraints
	Power system modeling
	State estimation
	Stability analysis
	Sensitivity analysis
	Reliability assessment
Planning, optimization, and forecasting	Optimization algorithms
	Economic dispatch
	Unit commitment
	Forecasting (load, generation, and weather)
Interconnected systems and market mechanisms	Interconnected power systems and tie-lines
	Grid codes and standards
	Regulatory frameworks and policies
	Market mechanisms and pricing

In the study of microgrid and power systems, several key terminologies form the foundation for understanding how electricity is generated, transmitted, and distributed. These terms include [11, 13]:

- Power system [3]: This refers to a complex network comprising various electrical components. The system encompasses the entire process of electricity production, its transmission through high-voltage lines, and distribution to end-users. It includes generation facilities like power plants, transmission infrastructure that carries high-voltage electricity over long distances, and distribution networks that deliver electricity to homes and businesses.
- Branch [14]: In the context of an electrical circuit, a branch represents any individual electrical element or a combination of elements that lies between two nodes. These elements can include components such as resistors, capacitors, inductors, or any other type of electrical device.
- Bus [15–18]: This term describes a central point in a power system where multiple electrical paths converge. In a power grid, a bus is typically a junction where various lines or components such as generators, transformers, or transmission lines are interconnected. It plays a crucial role in the distribution of electrical power within the network.
- Loop [19, 20]: A loop is defined as a closed-circuit path that starts and ends at the same node. It is a fundamental concept in circuit analysis, ensuring that the current flowing into a node also flows out, maintaining the principle of conservation of energy.
- Node [21, 22]: In electrical engineering, a node refers to a point where two or more circuit elements (such as wires) connect. At a node, all connecting points have the same electrical potential or voltage. Nodes are essential in analyzing and understanding complex electrical circuits, as they are the points where current either divides or combines.

In power systems, a branch represents any segment or component that connects two nodes, playing a crucial role in the network's functionality. Table 2 categorizes various types of branches, each with its distinct characteristics and purpose. From transmission lines that carry high-voltage electricity across long distances to transformers that adjust voltage levels for efficient distribution, each type of branch serves a specific function in the overall power transmission and distribution process. Understanding these branches is fundamental to comprehending how power systems are structured and how they operate to ensure continuous and reliable electricity supply.

Loops refer to the closed-circuit paths through which electrical current flows. Table 3 provides an overview of different types of loops, each contributing uniquely to the network's reliability and efficiency. From radial loops common in distribution networks to meshed networks in transmission systems, these loops offer various operational advantages, such as redundancy in case of faults and flexibility in power flow management. The design and configuration of these loops are critical for maintaining system stability and ensuring uninterrupted power supply in different scenarios.

Table 2 Classification and roles of branches in electrical power systems [14]

Type of branch	Description	Function/role
Transmission line	A high-voltage line for transferring electricity over long distances.	Transmits electricity from power plants to substations or between substations.
Distribution line	Lower voltage lines that distribute electricity to end-users.	Delivers electricity from substations to homes, businesses, and other consumers.
Feeder	A type of distribution line that originates at a distribution substation.	Provides electricity to a specific area or neighborhood.
Transformer branch	Connects a transformer to the network.	Steps up or steps down voltage levels for transmission or distribution.
Capacitor bank	Connected as a branch in power systems.	Improves power factor and voltage regulation, and reduces losses.
Reactor	An inductive component that is used to limit short-circuit currents and control voltage.	Regulates power flow and voltage levels, especially in high-voltage networks.

Table 3 Overview of different types of loops in power system networks [19, 20]

Type of loop	Description	Function/role
Radial loop	A simple loop system with one path from the source to the load.	Common in distribution networks, provides a single path for power flow.
Ring main	A circular loop where each component is connected to two other components.	Enhances reliability by providing alternate paths for power in case of a fault.
Meshed network	Multiple interconnections forming several loops.	High reliability and flexibility, commonly used in transmission networks.
Control loop	A loop in control systems for monitoring and regulating power systems.	Ensures system stability and efficiency through feedback and control mechanisms.

Nodes in power systems are junction points where electrical lines or components like generators and loads connect. Table 4 outlines the different types of nodes, highlighting their roles and functionalities within the electrical network. Nodes are pivotal in defining the structure of the network, whether they are generation nodes supplying power, load nodes where power is consumed, or substation nodes that facilitate power transformation and distribution. Understanding the types and functions of nodes is essential for analyzing and managing the complex dynamics of power systems.

The intricate network of a power system comprises various components crucial for the generation, transmission, and distribution of electricity. Among these components, buses hold significant importance as pivotal junction points. They facilitate the connection and interaction of different elements within the power grid, ranging from generators and transformers to transmission lines and loads. Table 5

Table 4 Types of nodes in power systems and their functions [21, 22]

Type of node	Description	Function/role
Generation node	A node where power generation units are connected.	Source of electricity supply in the network.
Load node	A node where loads (consumers) are connected.	Points of electricity consumption in the network.
Substation node	A node representing a substation in power systems.	Facilitates the transformation and distribution of power.
Interconnection node	Nodes where two or more power systems or networks meet.	Allows for the transfer of power between different networks or systems.
Ground node	A reference node representing the ground in electrical circuits.	Serves as a reference point for measuring voltages in the system.

Table 5 Types and functions of buses in power systems [15–18]

Type of bus	Description	Function/role
Generator bus	Also known as a PV (power-voltage) bus.	Connected to generators. Controls the voltage and the real power output of the connected generator.
Load bus	Also referred to as a PQ (power-quantity) bus.	Connected to loads (consumers of electricity). It is where the real and reactive power are specified.
Slack bus	Also known as a swing or reference bus.	Balances the active and reactive power in the system. Typically, one per system and usually a major generator with the ability to control voltage and frequency.
Interconnection bus	Connects two or more power systems or substations.	Facilitates power transfer between different systems or areas.
Distribution bus	Found in distribution systems.	Distributes power to various local loads. Often connected to transformers.
Substation bus	Located in substations.	A point of connection for incoming and outgoing transmission lines in a substation.
Ring bus	A configuration where each component is connected to two buses forming a ring.	Enhances reliability. If one bus fails, the other can handle the entire load.
Main bus and transfer bus	Used in substations with high reliability requirements.	Main bus carries normal load. Transfer bus is used to transfer load during maintenance or faults.
Auxiliary bus	Used to supply power to auxiliary systems like cooling and lubrication systems in power plants.	Ensures continuous operation of auxiliary equipment critical for the main equipment.

categorizes the different types of buses commonly encountered in power systems, each distinguished by its unique role and function. Understanding these buses is essential for grasping the operational dynamics of power systems and their efficiency in delivering electricity to end-users. From generator and load buses, which are directly connected to power generation and consumption, to specialized buses like slack and ring buses, each type plays a vital role in maintaining the stability and reliability of the power grid, which is extracted from the literature in Table 5 **[15–18]**.

2 Power Transfer

Power transfer is the process of moving electrical energy between generation sources, transmission lines, and distribution networks to end-users and maintaining system stability and reliability [13]. The concept of multidirectional power transfer applies to power system flow, which involves the potential difference between two points (nodes or branches) in the system. Active or reactive power, represented by P or Q, can be transferred from one node or branch to another. Consequently, the potential difference between two points in the system can be characterized by both magnitude and phase difference or expressed as having both magnitude and direction, shown in Figs. 1, 2, 3.

In power system analysis, the most frequently utilized quantities are phasors and scalar quantities [23]. A vector is a physical quantity characterized by both magnitude and direction, which can be represented in Cartesian coordinates (x, y, z) or polar coordinates. On the other hand, a phasor is a mathematical construct devised in the field of electronics to describe the behavior of alternating current (AC); it possesses both a magnitude and a phase, with the phase measured in degrees or radians. It is important to note that the phase of a phasor is distinct from the angular component in polar coordinates, as it specifically refers to the position of the waveform in its cycle relative to a reference.

The idea behind representing a system in space vector in terms of a phasor diagram is to illustrate the system's steady-state interrelationship between graphical quantities that vary sinusoidally in time. However, a phasor measures quantities mathematically in steady state; still, this method is widely applied for power systems and components (motors, drivers, transformers, etc.) and operation and

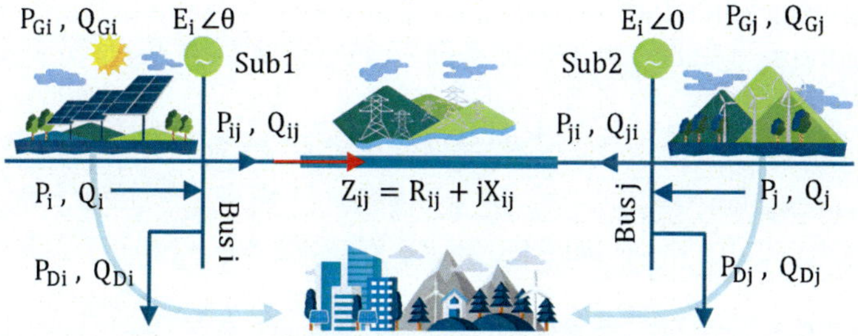

Fig. 1 Single-line diagram of power flow in an interconnection multisource power systemα: Angle between voltage E_i and $E_j\beta$: Angle of load or impedance angelϕ: Angle between voltage and current$i$ – subscript: Leading voltage $\angle\theta j$ – subscript: Lagging voltage $\angle 0 E_i \angle \theta$ to $E_j \angle 0$: Current flows from i to j

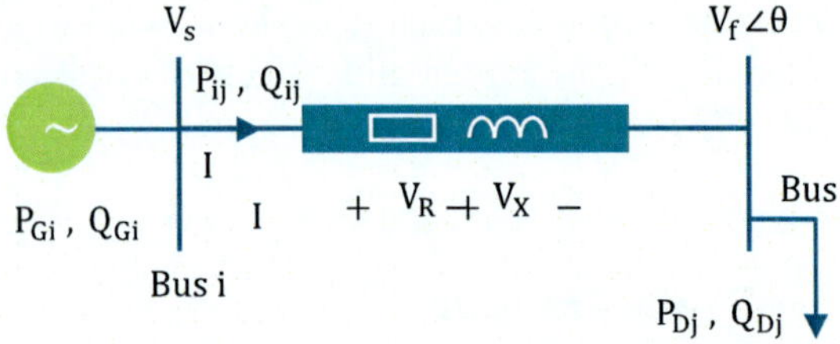

Fig. 2 Simplified representation of power transmission from generation to load with reactive support

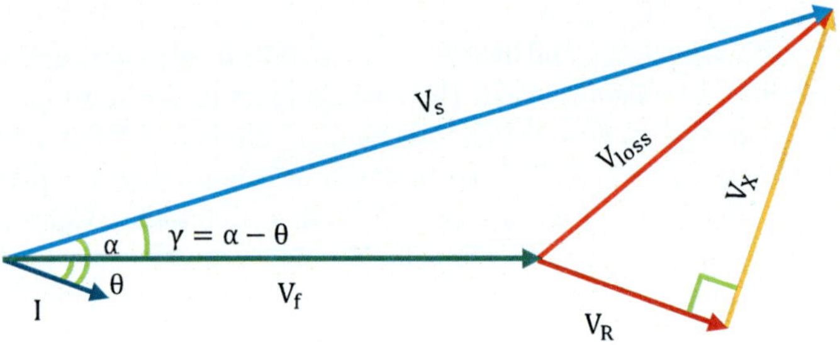

Fig. 3 Phasor diagram showing voltage and current relationships in an AC power system

behavior analysis [24, 25]. The system phasor diagram in the polar coordinate system is vectorized from a reference zero angle point of source voltage with an angle of AC source angle from the horizon direction:

$$\overline{V}_R = R\overline{I} \parallel \overline{I} \tag{1}$$

$$\overline{V}_X = jX\overline{I} \perp \overline{I} \tag{2}$$

The θ is the phase of the voltage with respect to the current I ($\measuredangle V_f - \measuredangle I$) called angle of the power factor (PF) of the load, and α is the power factor angle of the source (V_s). Cos θ is the load power factor that results in a big power factor angle for a smaller power factor and in a small power factor angle for a bigger power factor.

While γ is the phase difference between source and load (torque angel), which identifies with the torque (synchronous machine) to apply to the generator that is feeding the load, called the torque angle. I lagging V_f by θ angle: It is the slightly inductive load.

If $\theta = \measuredangle V_f - \measuredangle I > 0^{\circ}$, the load is observing positive reactive and active power as follows:

$$\cos\theta > 0 \Rightarrow P = V_f\, I\cos\theta > 0 \tag{3}$$

$$\sin\theta > 0 \Rightarrow Q = V_f\, I\sin\theta > 0 \tag{4}$$

If $\theta = \measuredangle V_f - \measuredangle I < 0^{\circ}$, the load is observing active power and generate reactive power as follows:

$$\cos\theta > 0 \Rightarrow P = V_f\, I \cos\theta > 0 \tag{5}$$

$$\sin\theta < 1 \Rightarrow Q = V_f\, I \sin\theta < 0 \tag{6}$$

If $\theta = \measuredangle V_f - \measuredangle I = 0^\circ$, the load is resistive and observes no reactive power, only observes active power as follows:

$$\cos\theta = 1 \Rightarrow P = V_f\, I \cos\theta > 0 \tag{7}$$

$$\sin\theta = 0 \Rightarrow Q = V_f\, I \sin\theta = 0 \tag{8}$$

If $\theta = \measuredangle V_f - \measuredangle I = 90^\circ$, the load is purely inductive and observes no active power and only observes reactive power as follows:

$$\cos\theta = 0 \Rightarrow P = V_f\, I \cos\theta = 0 \tag{9}$$

$$\sin\theta = 1 \Rightarrow Q = V_f\, I \sin\theta > 0 \tag{10}$$

If $\theta = \measuredangle V_f - \measuredangle I \Longrightarrow 180^\circ < \theta > 90^\circ$, the load delivers active power and observes reactive power as follows:

$$\cos\theta = 0 \Rightarrow P = V_f\, I \cos\theta = 0 \tag{11}$$

$$\sin\theta = 1 \Rightarrow Q = V_f\, I \sin\theta > 0 \tag{12}$$

A generalized conceptualization with respect to the load nature is given in the coordinate plane shown in Fig. 4.

According to the literature [11, 26, 27], in the context of power systems, the terms "magnitude" and "amplitude" are often used interchangeably, but they have distinct meanings, especially when discussing electrical signals such as voltage and current. Magnitude generally refers to the size or length of a vector in a complex plane. In power systems, when we talk about the magnitude of a voltage or current, we usually refer to its RMS (root mean square) value. The RMS value is a statistical measure of the magnitude of a varying quantity and is used to compare both alternating and direct currents (AC and DC) (or voltages). Essentially, it gives the equivalent direct current (or voltage) value that would produce the same power dissipation in a resistor. Amplitude, on the other hand, typically refers to the peak value of a sinusoidal waveform. It is the maximum extent of a vibration or oscillation, measured from the position of equilibrium. In the context of an AC (alternating current) power system, the amplitude would be the peak value of the voltage or current waveform.

Figure 5. provides a framework for an integrated approach to project management within the context of power system operations. The approach is systematic and encompasses a dual-faceted perspective: one that considers the technical flow of the system and another that focuses on the project management cycle. The feasibility

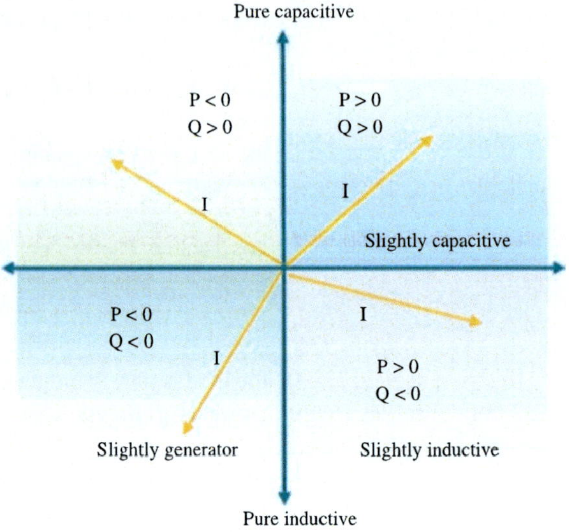

Fig. 4 Flow of active and reactive power in a typical system

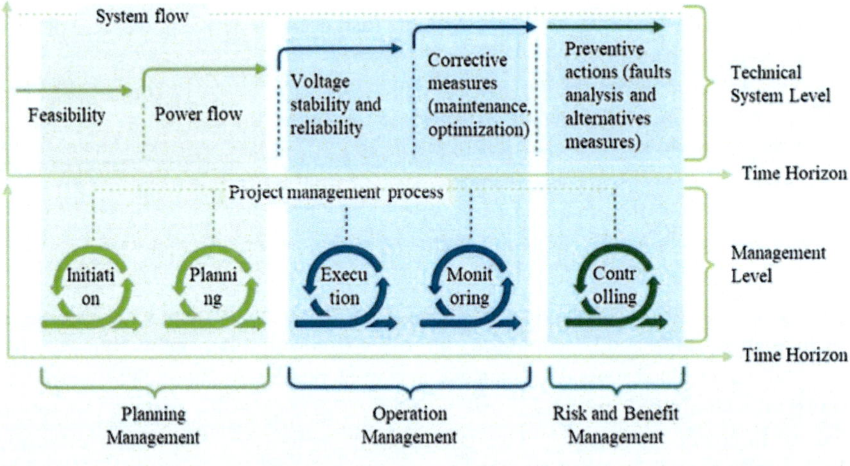

Fig. 5 Integrated project management and system operation framework in power systems

stage lays the groundwork for the power project by assessing the practical aspects and potential success, ensuring that the power flow is adequately planned and sustainable. This initial phase is closely tied to the initiation and planning stages of project management, wherein new projects are not only proposed but also meticulously prepared for.

Progressing along the system flow, attention shifts to "voltage stability and reliability," highlighting the demand for a power system to maintain consistent voltage levels and reliability throughout its operation. This technical requirement is in parallel with the "execution" phase of project management, where plans are brought to fruition and the system's design is implemented. Subsequent to the operationalization of the system, "corrective measures" are identified to optimize maintenance procedures, thus enhancing the operational efficiency and longevity of the power system. This corresponds with the "monitoring" and "controlling" phases of project management, where ongoing performance is evaluated against the benchmarks, and adjustments are made to steer the project toward its defined objectives.

Additionally, this figure shows the importance of "preventive actions," which align with the risk and benefit management aspect of the project management process. This stage is critical for conducting fault analysis and establishing contingency measures, which serve to preemptively mitigate potential issues before they materialize, thereby safeguarding the integrity of the power system. The integration of the system's technical flow with project management processes, encompassing initiation, planning, execution, monitoring, and controlling, establishes a robust strategy for managing power system projects. This strategy is not static but rather a dynamic and iterative cycle, requiring continuous reassessment and adaptation over time. The "time horizon" and "management level" further emphasize that each stage is influenced by temporal factors and decision-making hierarchies, which are pivotal in ensuring that the project aligns with both immediate and long-term objectives of power system stability and efficiency.

Figure 6 illustrates the interconnected processes of planning, operation, and control within power system management to ensure stability, reliability, and efficiency.

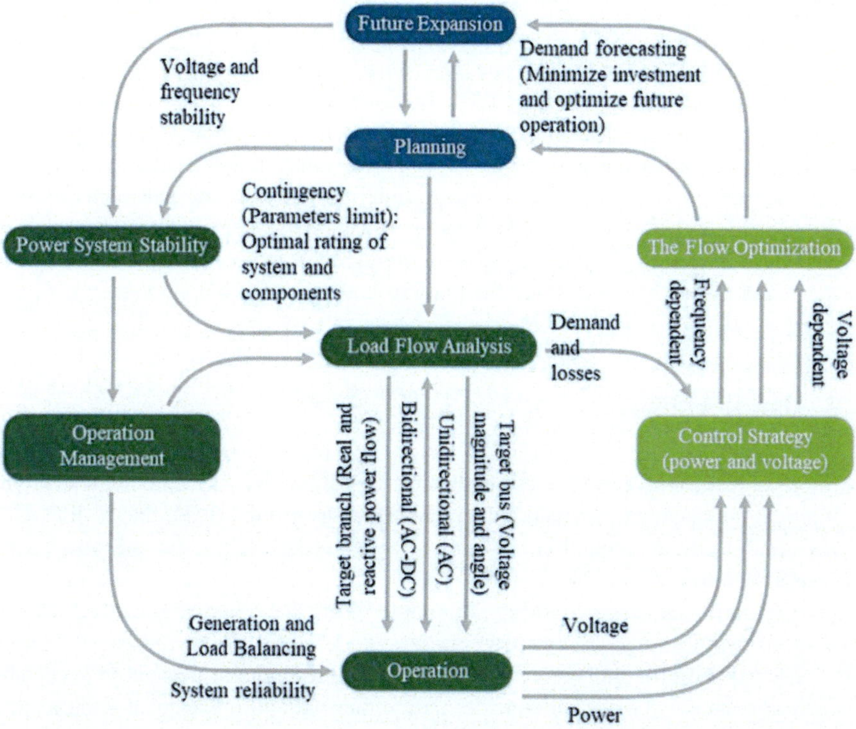

Fig. 6 Dynamic processes in power system management for stability and efficiency improvement

It emphasizes the cyclical nature of managing a power system, starting with "power system stability," which is foundational to the entire process. It involves ensuring voltage and frequency stability and setting parameters for system and component operations to handle contingencies. "Load flow analysis" is central to both planning for future expansion and day-to-day operation management. It assesses how power flows through the network, taking into account bidirectional flows in advanced systems and unidirectional flows in more traditional setups. This analysis is crucial for "the flow optimization," which looks to minimize losses and balance demand against generation.

In operation management, the focus is on "generation and load balancing" to maintain system reliability. This operational balancing act feeds into "control strategy," which adjusts power and voltage in real time to meet the fluctuating demands and maintain steady-state operation. The planning aspect considers "future expansion," utilizing demand forecasting to guide investment and future system operations. This is a forward-looking process that feeds back into the stability and operation cycles, ensuring the power system can adapt to future needs. It represents the continuous interaction between planning, analysis, and control in power system management to achieve a balance between current operations and future needs, ensuring that the system remains stable, reliable, and prepared for expansion.

Load flow analysis is a critical aspect of hybrid renewable energy systems, encompassing the modeling and evaluation of various combinations of renewable energy sources [27, 28]. This analysis is essential for optimizing system operations to ensure efficient energy distribution and reliability. By modifying the load flow algorithm, it is possible to calculate the critical voltage at each bus, which is pivotal for maintaining system stability [28]. Techniques such as the continuation power flow method and the utilization of the Jacobian matrix method are instrumental in this context, providing a rigorous approach to understanding and managing the complexities of renewable energy integration into power grids. These methods not only facilitate the assessment of system performance under different scenarios but also aid in the strategic planning of system upgrades and expansions, ensuring that the power system can withstand fluctuations and continue to operate effectively.

3 Fault Analysis

Short-circuit faults can be classified into two categories: asymmetrical and symmetrical faults. Asymmetrical faults include line-to-ground (LG), line-to-line (LL), and double-line-to-ground (LLG) faults. Symmetrical faults, on the other hand, consist of three-phase faults, such as triple line (LLL) and triple line to ground (LLLG) faults [29], shown in Fig. 7 (Adapted from [13, 29] and enhanced with the author's own ideas).

The capacity for short-circuit current, within the limits of maximum acceptable fault tolerance, is critical in considering the thermal and mechanical endurance of

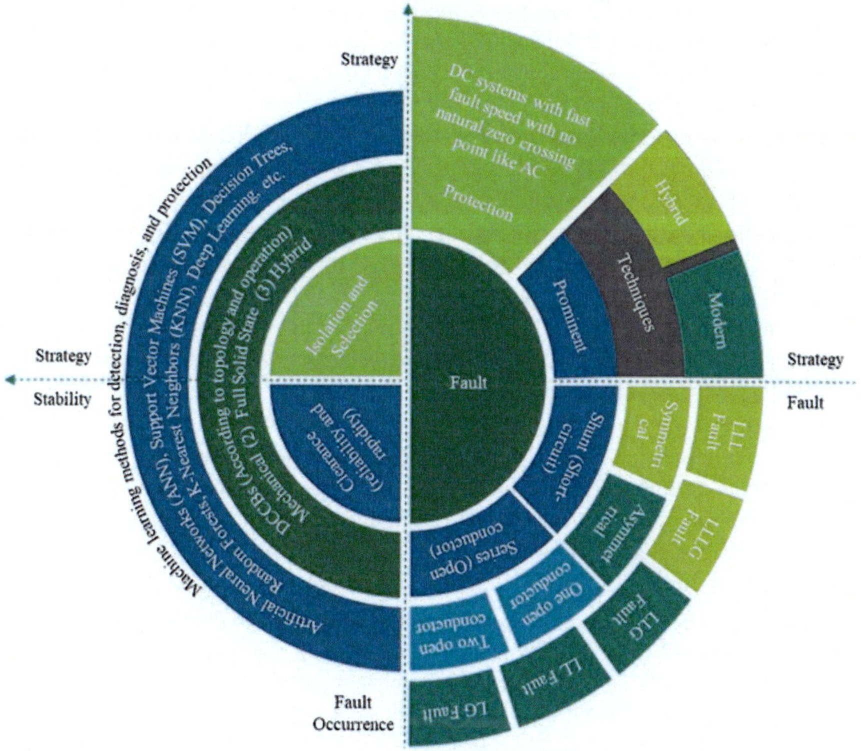

Fig. 7 Strategic framework for fault management in hybrid AC/DC microgrid systems

components. This study, suggested within the framework of IEC, ANSI/IEEE, and NEC standards, is supported by a recent literature review and comparative analysis. In the context of both upstream and downstream processes, the dominant source of short-circuit current must be meticulously understood. When integrating a Distributed Generation (DG) source into a power system, it is imperative to maintain harmony in voltage levels and frequency synchronization, along with addressing contingency and stability, power quality, and other factors. This integration necessitates a comprehensive calculation of total fault current, combining the main upstream grid and DG superpositions, to ensure safe and sustained operation [30].

Furthermore, the sizing of transformers in medium- and low-voltage radial networks is of significant importance. This is because the upstream grid's fault current capacity is largely dictated by the short-circuit impedance of the system's transformers [31].

Integrating distributed generators (DGs) into networks has operation challenges [32] despite offering flexible on-site power generation and reduced carbon emissions. One major concern is the impact on fault currents, which can be influenced by magnitude and direction [33]. Also, it increases fault current at the fault location but decreases it from the upstream utility grid, reducing the reach distance of the protection device [34]. Furthermore, synchronous generator-based DGs contribute more to fault current than converter-based DGs [35]. However, these findings are primarily based on balanced faults, whereas unbalanced earth faults may impact fault currents differently, particularly from converter-based DGs [36]. This study investigates the influence of a full converter-based DC source (equivalent) on fault current during both balanced and unbalanced faults, considering external grid short-circuit power capacity and DGs integration location. The study's results will improve understanding and management of DG integration's impact on distribution network operation.

Short-circuit analysis calculates the maximum available fault current in a system that can occur due to a fault [37]. This analysis depends on the current ratio, line parameters, or line impedance of the faulted branch and assesses the maximum level of available current that equipment can withstand. Asymmetrical short-circuit faults have two components: the AC component, which varies sinusoidally with time and is known as a symmetrical fault, and the DC component, which is nonperiodic and decays exponentially with time [37]. The DC component's decay time is proportional to the reactance (X) to resistance (R) ratio X/R, where L/R represents the inductance (L) to resistance (R) ratio. The inverse of X/R (1/(X/R)) equals the power factor (PF). The X/R ratio affects the DC component at the fault point, where a higher X/R ratio causes a longer time constant for the DC component to decay, resulting in a more prolonged graph. Conversely, a shorter X/R ratio results in a shorter time constant for the DC component to decay.

The X/R ratio determines the peak of an asymmetrical fault, which is used to select protection devices. Specifically, if the X/R ratio of a circuit is less than the X/R ratio of the circuit breaker being used, a symmetrical short-circuit current will be less than the symmetrical rating of the circuit breaker [38]. Therefore, selecting a circuit breaker with a suitable X/R ratio is crucial in ensuring the protection of the circuit against asymmetrical faults. The difference between the initial fault current and the final steady-state current depends on the supply voltage's X/R ratio and phase angle. At the start of the fault, the DC component is equal in magnitude but opposite in polarity to the AC component [38].

4 Optimization and Dataset Analysis

To minimize the size, cost, and emissions, the following objective functions for each renewable energy resource and storage system are considered (Table 6). For each renewable energy resource and storage system, the objective functions aim to minimize the system size (installed capacity), system cost, and CO_2 emissions. These objectives can be combined using multi-objective optimization techniques to find the optimal trade-offs between.

Table 6 Analysis of renewable energy resources: dataset components and optimization objectives

Resource	Dataset components	Objective functions
Solar PV plant (SPP)	Location (latitude, longitude, and elevation)	Minimize system size (installed capacity)
	Solar irradiance data (global horizontal irradiance, direct normal irradiance, diffuse horizontal irradiance)	Minimize system cost
	PV module specifications (efficiency, temperature coefficient, etc.)	Minimize CO_2 emissions
	Inverter specifications (efficiency, power rating, etc.)	
	System size (installed capacity)	
	Energy production data (hourly or sub-hourly)	
	Loss factors (soiling, shading, wiring, etc.)	
Wind power plant (WPP)	Location (latitude, longitude, and elevation)	Minimize system size (installed capacity)
	Wind speed data (hourly or sub-hourly)	Minimize system cost
	Wind direction data	Minimize CO_2 emissions
	Wind turbine specifications (power curve, hub height, rotor diameter, etc.)	
	System size (installed capacity)	
	Energy production data (hourly or sub-hourly)	
	Loss factors (turbine availability, wake losses, etc.)	
Battery energy storage (BES)	Battery type (Li-ion, lead-acid, etc.)	Minimize system size (installed capacity)
	Battery specifications (energy capacity, power rating, depth of discharge, round-trip efficiency, etc.)	Minimize system cost
	Charge/discharge profile data (hourly or sub-hourly)	Minimize CO_2 emissions
	State of charge data	
	System size (installed capacity)	
	Degradation factors (cycle life, calendar life, etc.)	
Hydrogen storage/ generation (HSG)	Electrolyzer specifications (efficiency, power rating, etc.)	Minimize system size (installed capacity)
	Fuel cell specifications (efficiency, power rating, etc.)	Minimize system cost
	Hydrogen storage specifications (type, storage capacity, pressure, etc.)	Minimize CO_2 emissions
	Hydrogen production and consumption data (hourly or sub-hourly)	
	System size (installed capacity)	
	Energy conversion and storage efficiency factors	
	Maintenance and operation data	

5 Case Study: Optimizing a Typical Hybrid Renewable Energy System

This case study presents a systematic procedure to optimize a hybrid renewable energy system model comprising a combination of solar photovoltaic (PV), wind, battery energy storage (BES), and hydrogen storage, summing up to a total installed capacity of 10 MW. The optimization aims to balance cost efficiency, carbon dioxide (CO_2) emissions, and system reliability. The system components are as follows:

- Solar PV: x_1 MW
- Wind: x_2 MW
- Battery energy storage (BES): x_3 MW
- Hydrogen storage: x_4 MW

Objective: Minimize system cost and CO_2 emissions while maximizing reliability, adhering to the constraint:

$$x_1 + x_2 + x_3 + x_4 = 10\,\text{MW} \tag{13}$$

Systematic Procedure for Optimization:

- Step 1: Establish decision variables: Define x_1, x_2, x_3, and x_4 as the capacities to be installed for solar PV, wind, BES, and hydrogen storage, respectively.
- Step 2: Define objective functions.
  - System cost: Calculate the total cost encompassing capital investment, operations, maintenance, and fuel for each energy source.
  - CO_2 emissions: Assess the emissions from each component. Account for manufacturing, installation, energy conversion, and electricity sources for charging.
  - Reliability: Gauge the consistency of energy supply by considering resource intermittency, availability, and storage effectiveness.
- Step 3: Formulate constraints.
  - Capacity constraint: Ensure the sum of all power sources equals the system's total capacity.
  - Specific constraints: Incorporate limits based on physical, environmental, and technological considerations.
- Step 4: Selection of optimization method: Choose a suitable algorithm, such as linear or nonlinear programming (LP or NLP) or evolutionary algorithms to handle the model's objectives and constraints.
- Step 5: Solve the model: Use appropriate software or a programming environment to compute the optimal configuration for x_1, x_2, x_3, and x_4.
- Step 6: Analysis of results: Interpret the optimal solution, examining the implications for cost, emissions, and reliability. Conduct sensitivity analyses to perceive how variations in inputs affect the outcomes.

Conclusion: Adopting this structured approach ensures that the proposed hybrid system not only meets its capacity target but does so in the most economically viable, environmentally friendly, and reliable manner possible. The methodology sets the stage for informed decision-making regarding the integration and management of renewable energy resources.

A surrogate model, also known as an approximation model or metamodel, is a simplified representation of a complex system or process [39]. It is used to estimate the behavior or performance of the original system without the need for computationally expensive simulations or evaluations. Surrogate models are often employed in optimization and design problems where evaluating the true objective function is time-consuming or costly.

In the context of optimization, surrogate models are trained to approximate the relationship between input variables and the objective function [40]. They are used to predict the performance of different solutions, allowing for a more efficient search of the solution space. Surrogate models can be constructed using various techniques, including [41]:

- Polynomial regression: Fitting a polynomial function to the data points to approximate the underlying relationship between input variables and the objective function.
- Neural networks: Using artificial neural networks to model complex, nonlinear relationships between input variables and the objective function.
- Gaussian processes: A probabilistic model that combines prior beliefs about the objective function with observed data points to make predictions and estimate uncertainty.
- Radial basis functions (RBF): A type of interpolation method that uses radial basis functions to approximate the objective function.
- Kriging: A geostatistical method that uses spatial correlation to model the objective function.

Surrogate models are widely used in optimization problems, especially in engineering design and multi-objective optimization, where evaluating the true objective function can be computationally expensive or resource-intensive. By using a surrogate model, the optimization process can be performed more efficiently, with fewer evaluations of the actual system or process.

Table 7 offers a structured guide for selecting an appropriate power flow analysis method based on the specific requirements of the system under study. It outlines the trade-offs between computational efficiency, accuracy, and ease of use, helping engineers and researchers choose the most suitable method for their analysis.

The analytic hierarchy process (AHP) [44–46] is applied to evaluate the feasibility of load flow methods the power system and their applications, establishing a set of criteria and sub-criteria to systematically compare and prioritize the alternatives.

From selection perspective:

- Analysis objective
  - Steady-state analysis.

Table 7 Comparative analysis of power flow analysis methods in electrical systems [4, 10, 27, 42, 43]

Criteria	Gauss–Seidel method	Direct current (DC) load flow	Newton-Raphson method	Fast decoupled load flow
System size and complexity	Suitable for small- to medium-sized systems due to simpler calculations and lower computational resource requirements.	Best suited for large systems with a high degree of sparsity and a focus on transmission level analysis.	Ideal for large and complex systems due to its robust algorithm that can handle nonlinearities effectively.	Appropriate for very large systems, designed to simplify the Jacobian matrix by decoupling it, which reduces complexity.
Convergence	Linear convergence; may require more iterations for larger systems or systems with higher R/X ratios.	Not applicable, as DC load flow assumes a linear system.	Quadratic convergence; typically requires fewer iterations than gauss–Seidel and is more likely to converge for nonlinear systems.	Fast convergence due to decoupling; fewer iterations needed compared to full Newton-Raphson.
Accuracy rate	Moderate accuracy; depends on the initial guess and system conditions.	Low accuracy; only suitable for approximating power flow in transmission systems without considering reactive power.	High accuracy; capable of precise calculations accounting for both active and reactive power.	High accuracy but may be slightly less than full Newton-Raphson due to the decoupling approximation.
Computational time	Less computationally demanding; however, it may be inefficient for very large systems.	Computationally efficient; simplifications reduce the amount of required calculations.	Computationally intensive; however, its faster convergence often compensates for the complexity.	Reduced computational time due to the simplified matrix and fewer iterations.
Tools availability	Widely available in many power system analysis tools due to its fundamental nature.	Common in power system stability analysis tools, particularly for initial system design and planning.	Extensively supported in advanced power system simulation software.	Available in most modern power system analysis tools, especially those designed for real-time simulations.
Ease of use and learning	Relatively easy to understand and implement, making it a good teaching tool for basic power flow concepts.	Straightforward for scenarios that can ignore reactive power, making it accessible for initial studies and estimations.	More complex due to its nonlinear nature, requiring a deeper understanding of power system dynamics.	More complex than gauss–Seidel but often easier than full Newton-Raphson due to its decoupled approach.

  - Dynamic analysis.
  - Fault analysis.
  - Optimal power flow.
  - Reliability assessment.
  - Power quality.

- Complexity.

  - System size (number of buses, lines, generators, etc.)
  - System topology (radial, meshed, etc.)
  - Degree of coupling between real and reactive power.
  - Presence of distributed energy resources (DERs).
  - Interconnected or islanded operation.
  - Control schemes (voltage control, frequency control, etc.)

- Accuracy rate.

  - Tolerance level for voltage, power, and current errors.
  - Convergence rate.
  - Impact of approximations or assumptions.
  - Model fidelity (representation of loads, generators, transformers, etc.)

- Computational time.

  - Simulation time.
  - Memory requirements.
  - Scalability for large systems.
  - Ease of parallelization.

- Software availability and compatibility.

  - Commercial or open-source software options.
  - Integration with other software tools.
  - Customizability and extensibility.
  - Technical support and user community.

- Ease of use and learning.

  - User interface and visualization capabilities.
  - Documentation and tutorials.
  - Required expertise and knowledge.

From application perspective:

- Steady-state analysis.

  - Power flow methods (Gauss–Seidel, Newton-Raphson, fast decoupled, etc.)
  - Balanced or unbalanced systems.
  - Three-phase or single-phase analysis.

- Dynamic analysis.
  - Time-domain simulations.
  - Frequency-domain analysis.
  - Stability assessment (transient, voltage, small-signal, etc.)
- Transient and fault analysis.
  - Symmetrical component method.
  - Short-circuit.
  - Sequence network analysis.
  - Protection coordination.
- Optimal power flow.
  - Linear programming (LP).
  - Nonlinear programming (NLP).
  - Interior-point methods.
  - Genetic algorithms and other evolutionary techniques.
- Reliability assessment.
  - Component-based reliability indices.
  - System-based reliability indices.
  - Monte Carlo simulation.
  - Voltage stability and reliability.
- Power quality.
  - Harmonic analysis.
  - Voltage sag and swell analysis.
  - Flicker analysis.

6 Discussion and Future Direction

The exploration of microgrid power flow analysis in the context of renewable energy integration, as presented in this study, reveals several critical insights and directions for future research. The intricate dynamics of microgrid systems, especially when incorporating renewable energy sources like solar and wind, presents unique challenges and opportunities for the field of power systems engineering. The continual evolution of technology in energy systems, particularly in renewable energy technologies and storage solutions, is a significant factor that will shape the future of microgrid power flow analysis. The increasing efficiency of photovoltaic cells, advancements in wind turbine design, and the rapid development of battery technologies are pivotal. Future research must focus on optimizing these technologies within microgrids, considering the variable nature of renewable energy sources. The integration of smart grid technologies, which allow for more efficient monitoring and control of power flows, is also an area ripe for exploration.

The study of power flow analysis has traditionally relied on established algorithms like Newton-Raphson and fast decoupled load flow. However, the complexity of microgrids, especially with the incorporation of renewable sources and bidirectional flows, demands more advanced and flexible algorithms. Future research should focus on developing and refining algorithms that can efficiently handle the stochastic nature of renewable energy sources and the complexity of modern microgrid systems. Machine learning and artificial intelligence offer promising avenues for creating predictive models and optimization algorithms that can adapt to changing conditions in real time. As microgrids become more prevalent, there is a growing need for comprehensive policy and regulatory frameworks that address the unique challenges and opportunities they present. Future research should explore the development of policies that encourage the adoption of microgrids and renewable energy sources while ensuring grid stability and reliability. This includes studies on tariff structures, incentives for renewable energy integration, and regulations that facilitate the interconnection of microgrids with the main grid.

The economic aspects of microgrids, particularly in terms of investment and operational costs, are crucial for their widespread adoption. Future research should delve into comprehensive economic analyses to understand the cost-benefit dynamics of microgrid implementation, especially when integrating various renewable energy sources. This includes exploring different business models for microgrid operation and ownership, which can make microgrids more financially viable and attractive to investors. An essential aspect of integrating renewable energy into microgrids is the environmental impact. Future research should continue to quantify and analyze the environmental benefits of microgrids, such as reductions in greenhouse gas emissions and fossil fuel dependence. This research should also explore the sustainability of microgrid systems, including the life cycle analysis of system components and the long-term environmental impacts of widespread microgrid adoption. The role of DERs in microgrids is a critical area for future research. This includes not only renewable energy sources but also energy storage systems, demand response programs, and electric vehicles. Research should focus on how these diverse resources can be effectively integrated and managed within a microgrid to enhance efficiency, reliability, and resilience. The resilience of microgrids, particularly in the face of natural disasters and other disruptions, is an increasingly important area of study. Future research should explore how microgrids can be designed and operated to enhance their resilience, ensuring continuous power supply during grid outages. This includes studying the impact of extreme weather events on microgrid components and developing strategies to mitigate these risks. As the energy landscape evolves, the interconnection of multiple microgrids and their scalability becomes a significant area for future research. This includes exploring the technical, economic, and regulatory challenges associated with connecting multiple microgrids and scaling up microgrid projects from small, community-level systems to larger, more complex networks. Finally, the human and social dimensions of microgrid adoption cannot be overlooked. Future research should consider the social, cultural, and behavioral aspects of microgrid implementation, including community engagement, education, and the social impacts of transitioning to decentralized energy systems.

7 Conclusion

This chapter explores the fundamental aspects of microgrid power flow analysis, with a special emphasis on the integration of renewable energy sources. Our investigation has highlighted the complexities inherent in microgrid systems, especially in the context of their evolving role within the broader electrical grid. The integration of renewable energy sources, such as solar and wind power, into microgrids presents both challenges and opportunities. These renewable sources introduce variability and uncertainty in power generation, necessitating advanced management and optimization techniques to ensure stability and efficiency. Our study underlines the importance of sophisticated control mechanisms, such as voltage source inverters and hybrid AC-DC systems, in effectively managing these challenges.

Furthermore, we have identified key areas for future research and development, including the need for advanced algorithms for power flow analysis, the development of supportive regulatory frameworks, and the exploration of sustainable economic models. The environmental impact and the role of distributed energy resources (DERs) in enhancing grid resilience and reliability have also been recognized as critical areas for future exploration.

The insights gained from this study not only contribute to the academic understanding of microgrid systems but also provide practical guidance for their design and operation. As the energy landscape continues to evolve, the continued exploration and development of microgrid technologies will play a crucial role in achieving a sustainable, resilient, and efficient energy in future.

References

1. Heshmati, A., Abolhosseini, S., Altmann, J.: The Development of Renewable Energy Sources and Its Significance for the Environment. Springer (2015)
2. Yaqobi, M.A., Matayoshi, H., Danish, M.S.S., Urasaki, N., Howlader, A.M., Senjyu, T.: Control and energy management strategy of standalone DC microgrid cluster using PV and battery storage for rural application. Int. J. Power Energy Res. **2**, 53–68 (2018). https://doi.org/10.22606/ijper.2018.24001
3. Danish, M.S.S., Senjyu, T., Funabashia, T., Ahmadi, M., Ibrahimi, A.M., Ohta, R., Rashid Howlader, H.O., Zaheb, H., Sabory, N.R., Sediqi, M.M.: A sustainable microgrid: a sustainability and management-oriented approach. Energy Procedia. **159**, 160–167 (2019). https://doi.org/10.1016/j.egypro.2018.12.045
4. Kumar, A., Ghose, T.: A Newton-Raphson-based unified load flow of grid-connected and islanded AC–DC microgrids. Int. Trans. Electr. Energy Syst. **31**, e13075 (2021). https://doi.org/10.1002/2050-7038.13075
5. Danish, S.M.S., Ahmadi, M., Danish, M.S.S., Mandal, P., Yona, A., Senjyu, T.: A coherent strategy for peak load shaving using energy storage systems. J. Energy Storage. **32**, 101823 (2020). https://doi.org/10.1016/j.est.2020.101823
6. Liu, B., Chen, Z., Yang, S., Wang, Y., Yang, K., Lu, C.: Primary frequency regulation scheme applicable to LCC – VSC series hybrid HVDC considering AC voltage stability at receiving end. Int. J. Electr. Power Energy Syst. **140**, 108071 (2022). https://doi.org/10.1016/j.ijepes.2022.108071

7. Yaqobi, M.A., Matayoshi, H., Danish, M.S.S., Lotfy, M.E., Howlader, A.M., Tomonobu, S.: Low-voltage solid-state DC breaker for fault protection applications in isolated DC microgrid cluster. Appl. Sci. **9**, 723–735 (2019). https://doi.org/10.3390/app9040723
8. IGBT Technologies and Applications Overview: How and When to Use an IGBT. Semiconductor Components Industries, LLC, Phoenix (2018)
9. Elsayed, A.T., Mohamed, A.A., Mohammed, O.A.: DC microgrids and distribution systems: an overview. Electr. Power Syst. Res. **119**, 407–417 (2015). https://doi.org/10.1016/j.epsr.2014.10.017
10. Feng, W., Yuan, C., Shi, Q., Dai, R., Liu, G., Wang, Z., Li, F.: Using virtual buses and optimal multipliers to converge the sequential AC/DC power flow under high load cases. Electr. Power Syst. Res. **177**, 106015 (2019). https://doi.org/10.1016/j.epsr.2019.106015
11. Danish, M.S.S.: Voltage Stability in Electric Power System: a Practical Introduction. Logos Verlag Berlin GmbH, Berlin (2015)
12. Rathor, S.K., Saxena, D.: Energy management system for smart grid: an overview and key issues. Int. J. Energy Res. **44**, 4067–4109 (2020). https://doi.org/10.1002/er.4883
13. Al Mtawa, Y., Haque, A., Halabi, T.: A review and taxonomy on fault analysis in transmission power systems. Computation. **10**, 144 (2022). https://doi.org/10.3390/computation10090144
14. Wang, Y., Xu, Q., Liu, M., Zheng, J.: A novel system operation mode with flexible bus type selection method in DC power systems. Int. J. Electr. Power Energy Syst. **103**, 1–11 (2018). https://doi.org/10.1016/j.ijepes.2018.05.007
15. Paschall, C., Levinson, L., Ojemann, J., Ko, A., Herron, J.: Data-driven spectral features of directional DBS electrodes and dDBS-ECoG connectivity. In: 2021 10th International IEEE/EMBS Conference on Neural Engineering (NER), pp. 566–569 (2021). https://doi.org/10.1109/NER49283.2021.9441109
16. Tang, J., Zhao, J., Zou, H., Ma, G., Wu, J., Jiang, X., Zhang, H.: Bus load forecasting method of power system based on VMD and bi-LSTM. Sustain. For. **13**, 10526 (2021). https://doi.org/10.3390/su131910526
17. Al-Hussein, A.-B.A., Tahir, F.R., Ouannas, A., Sun, T.-C., Jahanshahi, H., Aly, A.A.: Chaos suppressing in a three-buses power system using an adaptive synergetic control method. Electronics. **10**, 1532 (2021). https://doi.org/10.3390/electronics10131532
18. Jiang, T., Feng, Z., Wan, K., Li, Q.: A power flow adjustment strategy based on weak bus identification and new bus type conversion. Int. J. Electr. Power Energy Syst. **137**, 107705 (2022). https://doi.org/10.1016/j.ijepes.2021.107705
19. Hassan, F., Kumar, A., Pati, A.: Recent advances in phase locked loops for grid connected systems: a review. In: 2022 IEEE Delhi Section Conference (DELCON), pp. 1–6 (2022), https://doi.org/10.1109/DELCON54057.2022.9752810
20. Abubakr, H., Vasquez, J.C., Hassan Mohamed, T., Guerrero, J.M.: The concept of direct adaptive control for improving voltage and frequency regulation loops in several power system applications. Int. J. Electr. Power Energy Syst. **140**, 108068 (2022). https://doi.org/10.1016/j.ijepes.2022.108068
21. Gao, X., Ye, Y., Su, W., Chen, L.: Assessing the comprehensive importance of power grid nodes based on DEA. Int. J. Crit. Infrastruct. Prot. **42**, 100614 (2023). https://doi.org/10.1016/j.ijcip.2023.100614
22. Liu, Y., Song, A., Shan, X., Xue, Y., Jin, J.: Identifying critical nodes in power networks: a group-driven framework. Expert Syst. Appl. **196**, 116557 (2022). https://doi.org/10.1016/j.eswa.2022.116557
23. Zhang, P., Qi, C., Fang, H., Sun, X.: A semi-analytical approach for the flexural analysis of in-plane functionally graded magneto-electro-elastic plates. Compos. Struct. **250**, 112590 (2020). https://doi.org/10.1016/j.compstruct.2020.112590
24. Hughes, A., Drury, B.: Synchronous, permanent magnet and reluctance motors and drives. In: Electric Motors and Drives, pp. 307–373. Elsevier, Amsterdam (2019). https://doi.org/10.1016/B978-0-08-102615-1.00009-X

25. Bayliss, C.R., Hardy, B.J.: Power transformers. In: Transmission and Distribution Electrical Engineering, pp. 543–614. Elsevier, Amsterdam (2012). https://doi.org/10.1016/B978-0-08-096912-1.00014-9
26. Brucoli, M., Maione, B., Margarita, E., Torelli, F.: Sensitivity analysis in power system dynamic stability studies. Electr. Power Syst. Res. **4**, 59–66 (1981). https://doi.org/10.1016/0378-7796(81)90037-7
27. Danish, M.S.S., Sabory, N.R., Funabashi, T., Danish, S.M.S., Noorzad, A.S., Yona, A., Senjyu, T.: Comparative analysis of load flow calculation methods with considering the voltage stability constraints. In: 2016 IEEE International Conference on Power and Energy (PECon), pp. 250–255. IEEE, Melaka (2016). https://doi.org/10.1109/PECON.2016.7951568
28. Indu Rani, B., Saravana Ilango, G., Nagamani, C.: Power flow management algorithm for photovoltaic systems feeding DC/AC loads. Renew. Energy. **43**, 267–275 (2012). https://doi.org/10.1016/j.renene.2011.11.035
29. Prasad, A., Belwin Edward, J., Ravi, K.: A review on fault classification methodologies in power transmission systems: part – I. J. Electr. Syst. Inf. Technol. **5**, 48–60 (2018). https://doi.org/10.1016/j.jesit.2017.01.004
30. Papathanassiou, S.A.: A technical evaluation framework for the connection of DG to the distribution network. Electr. Power Syst. Res. **77**, 24–34 (2007). https://doi.org/10.1016/j.epsr.2006.01.009
31. Boutsika, T.N., Papathanassiou, S.A.: Short-circuit calculations in networks with distributed generation. Electr. Power Syst. Res. **78**, 1181–1191 (2008). https://doi.org/10.1016/j.epsr.2007.10.003
32. Coster, E.J., Myrzik, J.M.A., Kruimer, B., Kling, W.L.: Integration issues of distributed generation in distribution grids. In: Proceedings of the IEEE, pp. 28–39 (2011), https://doi.org/10.1109/JPROC.2010.2052776
33. Ustun, T.S., Ozansoy, C., Zayegh, A.: Recent developments in microgrids and example cases around the world – a review. Renew. Sust. Energ. Rev. **15**, 4030–4041 (2011). https://doi.org/10.1016/j.rser.2011.07.033
34. Nimpitiwan, N., Heydt, G.T., Ayyanar, R., Suryanarayanan, S.: Fault current contribution from synchronous machine and inverter based distributed generators. IEEE Trans. Power Deliv. **22**, 634–641 (2007). https://doi.org/10.1109/TPWRD.2006.881440
35. Walling, R.A., Saint, R., Dugan, R.C., Burke, J., Kojovic, L.A.: Summary of distributed resources impact on power delivery systems. IEEE Trans. Power Deliv. **23**, 1636–1644 (2008). https://doi.org/10.1109/TPWRD.2007.909115
36. Coffele, F., Booth, C., Dyśko, A., Burt, G.: Quantitative analysis of network protection blinding for systems incorporating distributed generation. IET Gener. Transm. Distrib. **6**, 1218–1224 (2012). https://doi.org/10.1049/iet-gtd.2012.0381
37. Li, J., Lu, Y., Lu, R.: Detection of early decay in navel oranges by structured-illumination reflectance imaging combined with image enhancement and segmentation. Postharvest Biol. Technol. **196**, 112162 (2023). https://doi.org/10.1016/j.postharvbio.2022.112162
38. Jiang, Z., Guo, C., Ye, C., Ding, Y., Tang, X., Fang, Y.: Design and implementation of an auto-oriented bidirectional ZCS hybrid circuit breaker. IEEE Access. **9**, 152421–152429 (2021). https://doi.org/10.1109/ACCESS.2021.3126313
39. Keshtegar, B., Bagheri, M., Fei, C.-W., Lu, C., Taylan, O., Thai, D.-K.: Multi-extremum-modified response basis model for nonlinear response prediction of dynamic turbine blisk. Eng. Comput. **38**, 1243–1254 (2022). https://doi.org/10.1007/s00366-020-01273-8
40. Danish, M.S.S., Nazari, Z., Senjyu, T.: AI-coherent data-driven forecasting model for a combined cycle power plant. Energy Convers. Manag. **286**, 117063 (2023). https://doi.org/10.1016/j.enconman.2023.117063
41. Misener, R., Biegler, L.: Formulating data-driven surrogate models for process optimization. Comput. Chem. Eng. **179**, 108411 (2023). https://doi.org/10.1016/j.compchemeng.2023.108411
42. Danish, M.S.S., Senjyu, T., Danish, S.M.S., Sabory, N.R., Narayanan, K., Mandal, P.: A recap of voltage stability indices in the past three decades. Energies. **12**, 1544 (2019). https://doi.org/10.3390/en12081544

43. Furukakoi, M., Danish, M.S.S., Howlader, A.M., Senjyu, T.: Voltage stability improvement of transmission systems using a novel shunt capacitor control. Int. J. Emerg. Electr. Power Syst. **19**, 1–12 (2018). https://doi.org/10.1515/ijeeps-2017-0112
44. Ramesh, B., Khedkar, M., Shahare, K., Chappa, S., Mitra, A.: Analytic hierarchy process-based optimal load scheduling framework in an islanded distribution network. Energy Rep. **9**, 519–523 (2023). https://doi.org/10.1016/j.egyr.2023.09.103
45. Islam, M.R., Aziz, M.T., Alauddin, M., Kader, Z., Islam, M.R.: Site suitability assessment for solar power plants in Bangladesh: a GIS-based analytical hierarchy process (AHP) and multi-criteria decision analysis (MCDA) approach. Renew. Energy. **220**, 119595 (2024). https://doi.org/10.1016/j.renene.2023.119595
46. Ilham, Z., Subramaniam, I., Jamaludin, A.A., Wan-Mohtar, W.A.A.Q.I., Halim-Lim, S.A., Ohgaki, H., Ishihara, K., Mansor, M.R.A.: Analysing dimensions and indicators to design energy education framework in Malaysia using the analytic hierarchy process (AHP). Energy Rep. **8**, 1013–1024 (2022). https://doi.org/10.1016/j.egyr.2022.07.126

Sustainable Energy Policies Formulation Through the Synergy of Backcasting and AI Approaches

Mir Sayed Shah Danish, Mikaeel Ahmadi, Hameedullah Zaheb, and Tomonobu Senjyu

1 Introduction

Backcasting is a policymaking approach that begins with a desired future outcome and works backward to identify the necessary steps and policies to achieve it. Unlike traditional forecasting, which relies on current trends and past experiences, backcasting focuses on creating sustainable and equitable energy, climate change, and the sustainability of urban planning for a long time [1]. This method encompasses several segments aligned with the envisioned future state. Backcasting models, tailored to a specific policy context, combine various elements to address the problem, considering available data, resources, and policymaker goals. The components of the backcasting approach in the context of energy policy development can be classified into four main groups based on their functional commonalities (Fig. 1) [1–3].

M. S. S. Danish (✉)
Energy Systems (Chubu Electric Power) Funded Research Division, Nagoya University, Nagoya, Japan
e-mail: danish@mdanish.me

M. Ahmadi
Research Promotion Unit, Co-Creation Management Department, University of the Ryukyus, Okinawa, Japan

H. Zaheb
Department of Energy Engineering, Faculty of Engineering, Kabul University, Kabul, Afghanistan

T. Senjyu
Department of Electrical and Electronic Engineering, University of the Ryukyus, Okinawa, Japan

M. S. S. Danish (ed.), *Unified Vision for a Sustainable Future*,
https://doi.org/10.1007/978-3-031-53574-1_6

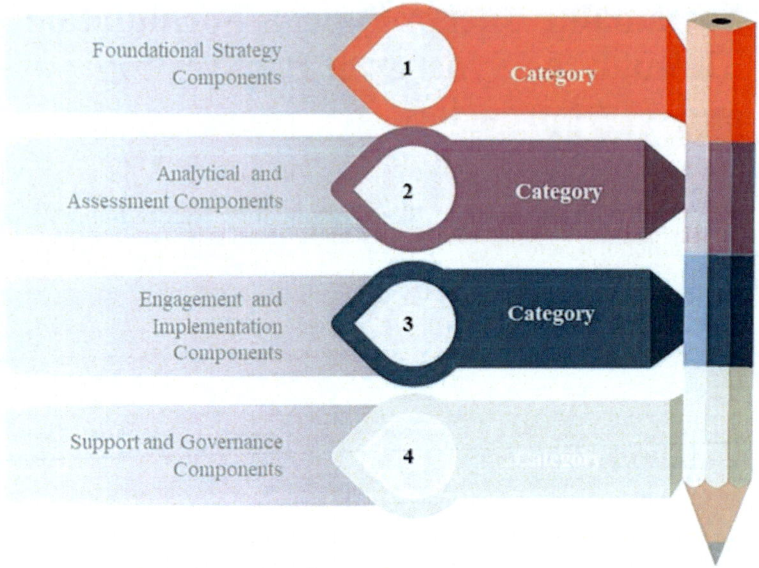

Fig. 1 Backcasting components in energy policy: foundational strategy, analytical and assessment, engagement and implementation, and support and governance components, illustrating the systematic process from vision to policy adaptation [1–3]

- Foundational strategy components
  - Defining the desired future state: Articulating the vision for the energy policy.
  - Scenario planning: Creating possible future scenarios to explore different pathways.
  - Systemic thinking: Integrating the energy policy within a larger system perspective.
  - Prioritization of scenarios: Evaluating scenarios to identify the most viable options.
  - Work backward from the future: Developing a reverse timeline from the future state to the present to identify necessary actions.
- Analytical and assessment components
  - Quantitative modeling: Utilizing mathematical models to forecast and evaluate policy impacts.
  - Feasibility analysis: Assessing the practicality and technical viability of policy scenarios.
  - Data collection and analysis: Gathering and examining relevant data to inform policy decisions.
- Engagement and implementation components
  - Stakeholder engagement: Involving relevant parties in the planning process.
  - Implementation planning: Creating action plans for enacting the energy policy.

    - Monitoring and evaluation: Continuously tracking the outcomes of implemented policies.
    - Adaptation and learning: Refining policies based on feedback and new information.
    - Communication and dissemination: Sharing policy-related information with stakeholders and the public.
    - Collaboration and partnership: Establishing alliances to support policy initiatives.
    - Public engagement and education: Involving the community and educating them on energy matters.

- Support and governance components

    - Flexibility and adaptability: Ensuring policies can evolve with changing circumstances.
    - Investment and financing: Securing funds for energy projects and initiatives.
    - Political will and leadership: Leadership advocating for and driving the energy policy agenda.

Backcasting models begin with defining a clear and measurable desired future state, which sets the foundation for the entire approach. This specification ensures that everyone involved understands the ultimate goal. Stakeholder engagement plays a crucial role in the backcasting process. By involving a diverse range of stakeholders, it is possible to build support for the desired future state and the policies needed to achieve it. Collaboration with stakeholders can also lead to more effective and inclusive policy solutions [4]. Scenario planning is another vital element of backcasting [1]. By developing different scenarios, policymakers can explore the feasibility and consequences of various policy choices. This process helps identify the most promising options and anticipate potential challenges.

Systemic thinking enables a holistic understanding of the policy problem [5]. A systems-based approach considers the interconnectedness of different elements within the policy issue, ensuring that all relevant factors are taken into account. Quantitative modeling helps estimate the impacts of different policies and scenarios. Mathematical models provide insights into the potential consequences of each option, assisting decision-makers in choosing the most effective path. Feasibility analysis evaluates different scenarios and policies based on the current state of technology, resources, and political will. This step ensures that selected strategies are realistic and achievable, given existing constraints. Once various scenarios have been assessed, prioritization of the most promising and feasible options occurs. This step helps focus resources and efforts on the strategies most likely to succeed.

Working backward from the future involves identifying the intermediate steps and policies necessary to achieve the desired future state [6]. This process helps break down the overarching goal into smaller and more manageable tasks. Implementation planning involves developing detailed plans to implement policies and programs to achieve the desired future state. This step is crucial for ensuring that the necessary resources, timelines, and milestones are in place. Monitoring and evaluation are essential to track the progress toward the desired future state. Regular assessments allow for adjustments to policies and programs as needed, ensuring the approach remains on track [7]. Adaptation and learning involve continuously refining policies and programs based on experience. This ongoing process ensures that the strategies remain effective and relevant in the face of changing circumstances.

Communication and dissemination play a vital role in sharing information about the backcasting approach, the desired future state, and the policies needed to achieve it [1]. By informing the public, policymakers, and stakeholders, support for the initiative can be built and maintained. Collaboration and partnership involve working with relevant organizations and stakeholders to develop and implement policies and programs that support the desired future state. These relationships can lead to more effective and efficient outcomes. Flexibility and adaptability are crucial when responding to new information and circumstances that arise during the backcasting process. Being open to change ensures that the approach remains responsive and dynamic [7]. Data collection and analysis inform decision-making, assess progress, and refine policies and programs as necessary. Using data-driven insights, the backcasting approach can be more effective and targeted.

Investment and financing involve securing funding and resources to support the implementation of policies and programs that will help achieve the desired future state [7]. Adequate financial support is crucial to the success of any initiative. Political will and leadership are necessary for the successful implementation of policies and programs. Strong leadership can help overcome obstacles, build support, and ensure that the desired future state is achieved [4]. Finally, public engagement and education are vital for building support for the desired future state and the policies needed to achieve it. By involving the public and promoting awareness, a broader base of support can be established, increasing the likelihood of success.

The illustration of the "Backcasting Classification" system through a structured flowchart in Fig. 2, segmented into four primary sections, each representing a core aspect of the backcasting process. The first section, "Vision and Planning," branches out into specific activities such as establishing visions, goals, and objectives, alongside recognizing critical trends and drivers that may impact future scenarios. The second section, "Analysis and Evaluation," is dedicated to scrutinizing strategies and their outcomes, suggesting a need for flexibility and continuous learning within the planning process. Moving on, the "Collaboration and Engagement" section underscores the importance of collective involvement in the backcasting methodology, highlighting the roles of stakeholder engagement, collaborative strategy formulation, and effective communication. Finally, the "Implementation and Support" section addresses the practical execution of the backcasting plan, emphasizing the necessity for supportive structures, be they financial, institutional, or educational, to realize the strategies laid out. Each section is interconnected, denoting a logical progression from conceptualization to realization, which is fundamental in strategic foresight and planning disciplines. This visual representation serves as a comprehensive guide to the various components and considerations that underpin the backcasting approach to planning and decision-making.

Table 1 shows a matrix of the interconnected nature of the backcasting approach, with "strong" and "medium" indicating the level of the interrelationship between

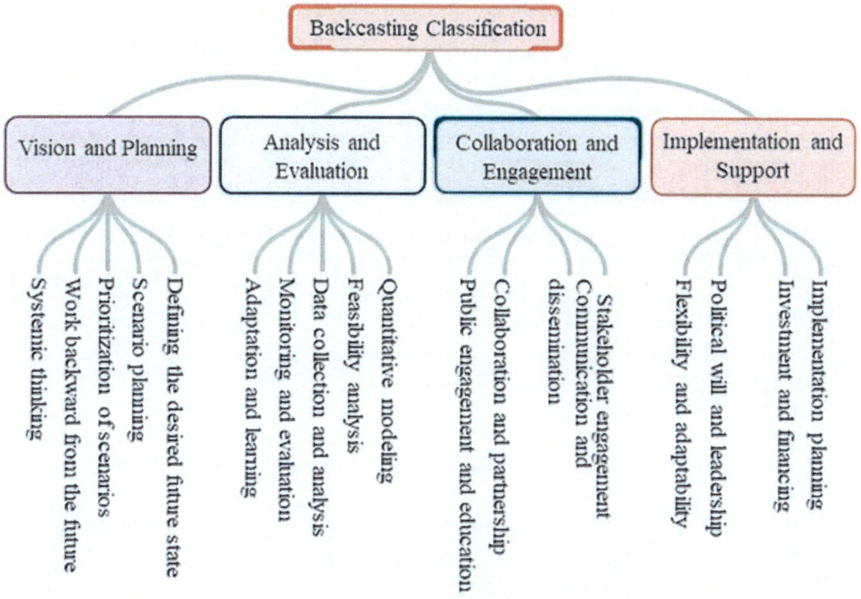

Figure 2 Structured flowchart of backcasting classification, depicting the four primary stages – vision and planning, analysis and evaluation, collaboration and engagement, and implementation and support – in strategic planning and policy implementation

the components and categories, emphasizing how each component influences and supports others.

2 Implementation Roadmap for Backcasting in Energy Policy: A Case Study Approach

An example implementation roadmap for the backcasting approach demonstrates its practical application in the field. By following these steps, policymakers can develop and implement a comprehensive energy policy that aligns with the Sustainable Development Goals (SDGs) and contributes to a more sustainable and equitable future.

- Defining the desired future state: Set a goal to achieve 100% renewable energy and reduce greenhouse gas emissions by 80% by 2050, in alignment with the SDGs.
- Stakeholder engagement: Bring together a diverse group of stakeholders, including government agencies, utility companies, environmental organizations, industry representatives, and local communities, to discuss and agree on the desired future state.

Table 1 Strategic integration of backcasting components in energy policy development [1, 8]

	Vision and planning	Analysis and evaluation	Collaboration and engagement	Implementation and support
Defining the desired future state	–	Medium	Strong	Medium
Scenario planning	Strong	Strong	Strong	Medium
Prioritization of scenarios	Strong	Strong	Medium	Strong
Work backward from the future	Strong	Strong	Medium	Strong
Systemic thinking	Strong	Strong	Medium	Medium
Quantitative modeling	Medium	–	Medium	Strong
Feasibility analysis	Strong	Strong	Medium	Strong
Data collection and analysis	Medium	Strong	Medium	Strong
Monitoring and evaluation	Medium	Strong	Strong	Strong
Adaptation and learning	Medium	Strong	Strong	Strong
Stakeholder engagement	Strong	Medium	–	Strong
Communication and dissemination	Medium	Medium	Strong	Strong
Collaboration and partnership	Medium	Medium	Strong	Strong
Public engagement and education	Medium	Medium	Strong	Medium
Implementation planning	Medium	Strong	Strong	–
Investment and financing	Medium	Medium	Strong	Strong
Political will and leadership	Medium	Medium	Strong	Strong
Flexibility and adaptability	Medium	Strong	Medium	Strong

- Scenario planning: Develop multiple scenarios representing different pathways to achieving the goal, including various technology options, policy interventions, and investment strategies.
- Systemic thinking: Consider the interconnections between energy systems, transportation, land use, and economic development in each scenario.
- Quantitative modeling: Use mathematical models to estimate the impacts of each scenario on energy supply, demand, emissions, and cost.
- Feasibility analysis: Evaluate the feasibility of each scenario based on current technology, resource availability, and political will.
- Prioritization of scenarios: Select the most promising and feasible scenario(s) to pursue.

- Work backward from the future: Identify intermediate milestones and policy measures necessary to achieve the desired future state, as outlined in the selected scenario(s).
- Implementation planning: Develop detailed plans for implementing policies and programs, such as renewable energy incentives, carbon pricing, energy efficiency standards, and infrastructure investments.
- Monitoring and evaluation: Regularly track progress toward the desired future state and adjust policies and programs to stay on course.
- Adaptation and learning: Continuously learn from experience, best practices, and emerging technologies and make adaptations as necessary to ensure that policies and programs remain practical and relevant.
- Communication and dissemination: Share information about the backcasting approach, the desired future state, and the policies and programs necessary to achieve it with a broader audience, including the public, policymakers, and stakeholders.
- Collaboration and partnership: Build partnerships with relevant organizations and stakeholders to develop and implement policies and programs that support the desired future state.
- Flexibility and adaptability: Be open to change and adapt to new information and circumstances as they arise.
- Data collection and analysis: Collect and analyze data to inform decision-making, evaluate progress, and refine policies and programs as necessary.
- Investment and financing: Secure funding and other resources to support the implementation of policies and programs that will help achieve the desired future state.
- Political will and leadership: Build political will and leadership to support the implementation of policies and programs necessary to achieve the desired future state.
- Public engagement and education: Engage the public and promote education and awareness to build support for the desired future state and the policies necessary to achieve it.

This case study design, through its comprehensive implementation roadmap, exemplifies the pragmatic application of the backcasting approach in energy policy development. The roadmap effectively integrates a series of steps, from setting ambitious renewable energy goals to stakeholder engagement, scenario planning, and systemic thinking. It further includes quantitative modeling, feasibility analysis, and the prioritization of scenarios, culminating in detailed implementation planning. Essential to this process are the aspects of monitoring, evaluation, adaptation, and learning, ensuring that the policy remains dynamic and responsive to new information and challenges. Key to the success of this roadmap is the emphasis on collaboration, communication, and public engagement, coupled with a commitment to flexibility and adaptability. By focusing on data-driven decision-making and securing necessary investment and political support, this case study highlights how a well-structured backcasting approach can lead to tangible, sustainable outcomes aligned with the SDGs.

3 The Assess, Strategize, Harmonize, Execute, and Sustain (ASHES) Framework

Adopting the Assess, Strategize, Harmonize, Execute, and Sustain (ASHES) framework in energy policy development can result in a comprehensive analysis of the five key domains as shown in Fig. 3 [9].

According to the literature [8, 10–12], this concept can be defined as follows. The access stage involves evaluating the current energy landscape, including energy supply and demand, infrastructure and technology, and environmental impact and emissions. Additionally, opportunities for artificial intelligence (AI) in energy policy are identified, such as predictive maintenance, demand-side management, renewable energy integration, energy storage, and electric vehicle infrastructure. The strategize stage introduces goals and objectives for AI integration in energy policy, leading to the development of a detailed AI deployment roadmap. This involves prioritizing AI applications, identifying required resources, and engaging key stakeholders like government agencies, utility companies, and technology providers. The harmonize stage addresses that AI-driven energy policies are aligned with existing regulations, focusing on creating a supportive regulatory environment for AI innovation. Collaboration between AI and energy experts is fostered, and ethical and societal concerns are addressed. Finally, execute deals with the implementation of AI-driven policies includes deploying AI technologies for efficient grid management and supporting research and development (R&D) in AI and energy sectors. Monitoring and evaluation are crucial, with performance indicators and data analysis to adjust policies as necessary.

3.1 Assess

- Assess the current energy landscape
    1. Energy supply and demand
    2. Infrastructure and technology
    3. Environmental impact and emissions

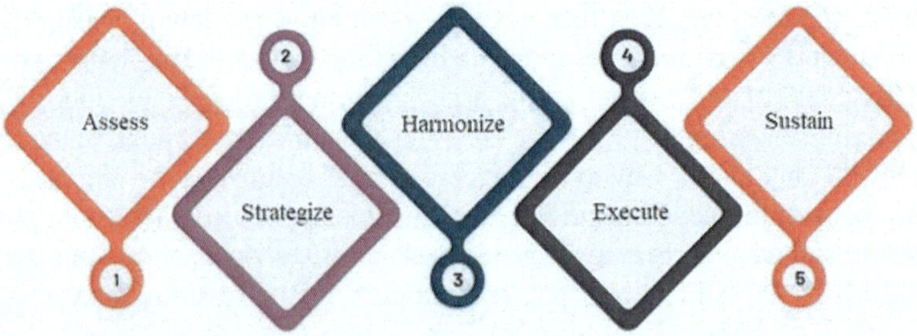

Figure 3 The ASHES framework in energy policy development, depicting the five key stages – assess, strategize, harmonize, execute, and sustain – for a comprehensive approach to energy policy analysis

- Identify opportunities for AI in energy policy
  1. Predictive maintenance and optimization
  2. Demand-side management and energy efficiency
  3. Integration of renewable energy sources
  4. Energy storage and grid management
  5. Electric vehicle charging infrastructure

3.2 *Strategize*

- Define goals and objectives for AI integration in energy policy
- Develop a roadmap for AI deployment
  1. Prioritize AI applications based on impact and feasibility
  2. Identify necessary resources and investments
  3. Establish a timeline for implementation
- Engage stakeholders
  1. Government agencies
  2. Utility companies
  3. Technology providers
  4. Research institutions
  5. Public and private sector organizations

3.3 *Harmonize*

- Align AI-driven energy policies with existing regulations and standards
- Establish a supportive regulatory framework for AI innovation in the energy sector
- Foster collaboration between AI and energy experts
- Address ethical and societal concerns related to AI in energy policy

3.4 *Execute*

- Implement AI-driven energy policies and programs
  1. Deploy AI technologies for grid management, energy efficiency, and more
  2. Support research and development in AI and energy
  3. Encourage private sector investments in AI-driven energy solutions
- Monitor progress and evaluate the effectiveness of AI interventions
  1. Establish performance indicators
  2. Collect and analyze data
  3. Adjust policies and programs as needed

3.5 *Sustain*

- Promote continuous learning and improvement in AI-driven energy policy
- Encourage knowledge sharing and collaboration among stakeholders
- Build public awareness and support for AI integration in energy policy
- Foster a culture of innovation and adaptability in the energy sector

Applying the ASHES framework in energy policy development through argumentation and analysis examines the application of the framework in developing a national energy policy. The focus is on a hypothetical scenario where a country aims to achieve a 50% renewable energy share and a 40% reduction in CO_2 emissions by 2030. The discussion and arguments are based on theoretical and empirical considerations.

3.6 *Assess*

- Current energy landscape analysis
  - Energy supply: Currently, 80% from fossil fuels and 20% from renewables. The argument here is for a transition toward more sustainable sources, considering the environmental and economic implications.
  - Energy demand: At 300 TWh per year, growing at 2% annually, it presents a challenge in balancing growth with sustainable practices.
  - CO_2 emissions: With 200 million tons per year, a significant reduction is imperative for environmental sustainability.
- AI integration opportunities
  - Predictive maintenance: Reduction in operational costs and increased efficiency of energy infrastructure.
  - Smart grid management: Enhanced load management, leading to energy savings and grid stability.
  - Renewable energy integration: Improved efficiency and management of intermittent renewable sources.

3.7 *Strategize*

- AI integration goals
  - Achieve a renewable energy mix of 50% by 2030.
  - Reduce CO_2 emissions by 40% to mitigate climate impact.
  - Enhance grid efficiency through AI technologies.

- Roadmap for AI deployment
  - Prioritize AI applications that offer the highest impact on sustainability and feasibility.
  - Budget allocation: An estimated $500 million investment over 10 years.
  - Timeline: A 10-year phased approach, with periodic evaluations.
- Stakeholder engagement
  - Involving government, industry, academia, and civil society to ensure multi-faceted perspectives and buy-in.

3.8 Harmonize

- Policy and regulation
  - Align AI-driven energy policies with existing regulations and global standards.
  - Create a supportive regulatory environment for AI adoption in the energy sector.
- Ethical and societal considerations
  - Address data privacy, security, and the societal impact of AI in energy.

3.9 Execute

- Implementation of AI policies
  - Deploy AI for grid management, renewable integration, and energy efficiency.
  - Support R&D in AI for energy innovations.
  - Incentivize private sector investment in AI for energy solutions.
- Monitoring and adaptive management
  - Track renewable energy share, emission reductions, and grid performance.
  - Use data analytics for informed decision-making and policy adjustments.

3.10 Sustain

- Long-term policy evolution
  - Promote continuous learning and improvement in AI applications within energy policy.
  - Encourage stakeholder collaboration for evolving knowledge and practices.

– Public engagement
    - Build awareness and support for AI in energy, emphasizing its role in sustainability.

The ASHES framework, when applied through rigorous argumentation and analytical methods, provides a structured approach to develop a sustainable and AI-integrated energy policy. The hypothetical scenario demonstrates the feasibility and strategic importance of this approach in achieving substantial improvements in renewable energy adoption and emission reductions while also addressing the socioeconomic and ethical dimensions of such a transformative policy.

Analyzing the ASHES framework in the context of energy policy development reveals a comprehensive, multifaceted approach to address current and future energy challenges. At the same time, the ASHES refers to audience, scope, hierarchy, explanation, and signposts. Within this terminology, the concept of ASHES becomes straightforward and more obvious. The framework provides a structured methodology for policymakers to assess, strategize, harmonize, execute, and sustain energy policies effectively [13].

The assess stage is crucial for understanding the current energy landscape. By evaluating the existing energy supply and demand, along with the infrastructure, policymakers can identify areas that need immediate attention and improvement. The high rating for "projected energy demand and supply scenarios" underscores the importance of forward-looking planning [14]. This foresight is critical in anticipating future energy needs and preparing adequately. The assessment also involves understanding how demographic, economic, technological, and environmental factors interplay with energy demand and supply, ensuring a holistic approach to energy policy formulation [15]:

– Current energy needs and trends
– Projected energy demand and supply scenarios
– Demographic, economic, technological, and environmental factors affecting energy demand and supply
– Energy infrastructure assessment
– Energy pricing and market analysis

The strategize phase focuses on setting clear objectives and developing a robust strategy for energy policy. The highest-rated aspects here are the setting of energy policy objectives and targets and the monitoring of progress through key performance indicators. This underscores the need for well-defined goals and measurable outcomes in energy policy. Additionally, considering different energy policy options and integrating AI and digital technologies into the strategy are essential for modern, efficient, and effective energy management. The integration of AI, although rated lower, is a forward-thinking approach that reflects the growing importance of technology in energy management [14, 15]:

– Energy policy objectives and targets set in each category
– Time-bound milestones for achieving the objectives and targets
– Key performance indicators for monitoring progress

- Energy policy options considered
- Weightings given to each option in the energy policy strategy
- Trade-offs and considerations involved in decision-making
- Scenario planning and modeling
- Integration of AI and digital technologies into energy policy

In the harmonize stage, the emphasis is on aligning the proposed energy policies with national and international regulations [16]. The high ratings for "consistency with national development plans" and "compliance with international agreements" highlight the importance of aligning energy policies with broader development goals and global standards. This ensures that the energy policies are not only locally relevant but also globally responsible [16]:

- Consistency with national development plans and strategies
- Compliance with international agreements and standards
- Coordination with regional and local energy policies

The execute phase is where policies are put into action. The high ratings for "performance of the policy against the targets set" and "progress of energy policy implementation" indicate that monitoring and evaluation are critical for the success of energy policies [14]. Regularly assessing the impact of these policies and making necessary adjustments ensures they remain effective and relevant. The role of AI in policy implementation, while rated lower, is an emerging area that can significantly enhance policy effectiveness through data-driven insights and automation [12, 17]:

- Progress of energy policy implementation
- Performance of the policy against the targets set
- Modifications made over time to improve policy effectiveness
- Key performance indicators for monitoring progress
- Role of AI in policy implementation and monitoring

Finally, the sustain phase emphasizes stakeholder engagement and continuous improvement [16, 17]. The high ratings for stakeholder engagement aspects reflect the recognition that energy policy development is not just a governmental task but requires the involvement of various groups, including the public, private sector, and civil society. Public awareness and education, although rated lower, are crucial for garnering public support and ensuring the long-term sustainability of energy policies [16, 17]:

- Level of stakeholder engagement in the energy policy development process
- Different groups and the extent of their participation and feedback
- Key stakeholders and their influence on the energy policy development process
- Public awareness and education campaigns
- Inclusive and participatory decision-making

The ASHES framework offers a comprehensive approach to develop and implement energy policies. It underscores the importance of a holistic view, considering not only the immediate energy needs but also the future scenarios, technological advancements, and the need for alignment with broader economic and environmental objectives [12]. The framework also emphasizes the critical role of continuous monitoring, stakeholder engagement, and adaptability, ensuring that energy policies remain effective and relevant in a rapidly changing world. The proposed road map presented in [4, 18–20] enables policymakers to systematically develop and implement energy policies that leverage AI technologies to drive sustainable, efficient, and resilient energy systems. It is worth mentioning that the importance rankings are subjective and may vary depending on the specific context and priorities of the policy development process.

Numerical assumptions in Figs. 4, 5, 6, 7, and 8 represent an exhaustive analysis and decision-making framework for energy policy development. Figure 4 assesses energy demand and supply, rating environmental factors as most impactful. Figure 5 prioritizes policy objectives, with fossil fuel reduction being the highest. Figure 6 demonstrates strategy development, emphasizing trade-offs as crucial in decision-making. Figure 7 tracks policy implementation, with modifications for effectiveness rated highest. Finally, Fig. 8 evaluates stakeholder engagement, with the extent of participation and feedback considered paramount.

The values assigned to various factors in the above information provide their relative importance or impact in the context of energy policy development. As the energy landscape evolves due to technological advances, economic shifts, regulatory changes, or environmental considerations, these values may need to be adjusted to accurately reflect the current state of affairs. For example, if new technological innovations emerge that significantly enhance energy efficiency, the importance rating assigned to "technological factors affecting energy demand" may increase. Conversely, if demographic changes slow down, their impact on energy supply might be deemed less critical, leading to a lower importance rating. Updating these

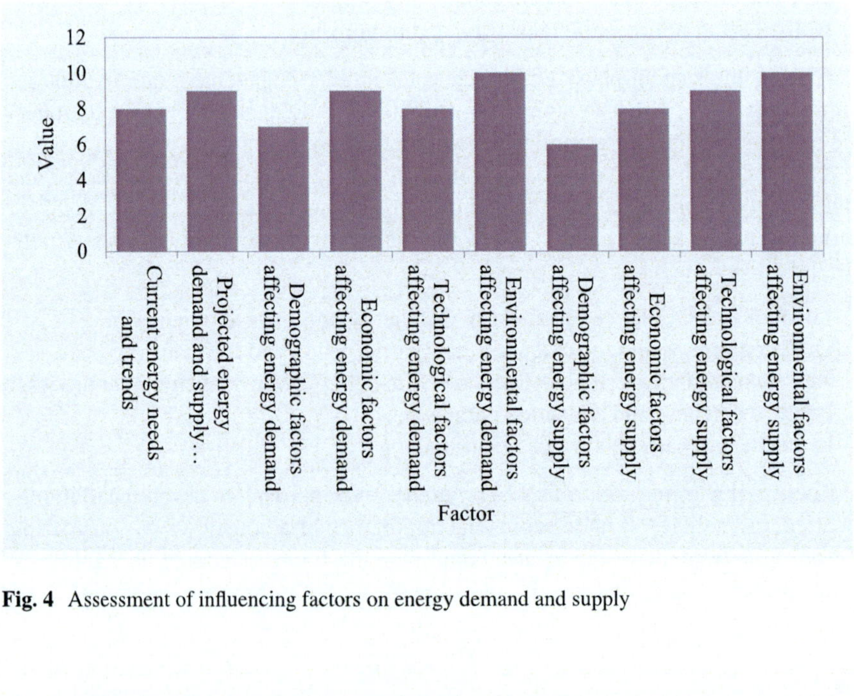

Fig. 4 Assessment of influencing factors on energy demand and supply

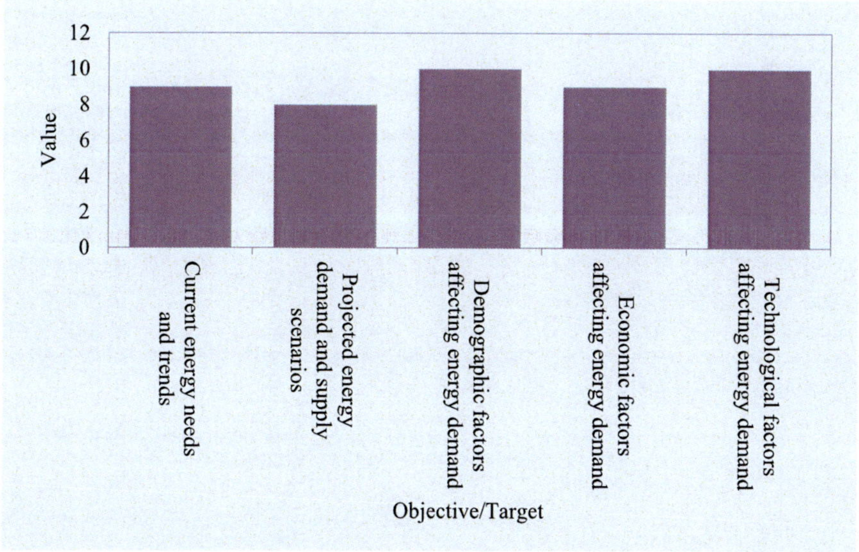

Fig. 5 Prioritization of energy policy objectives and targets

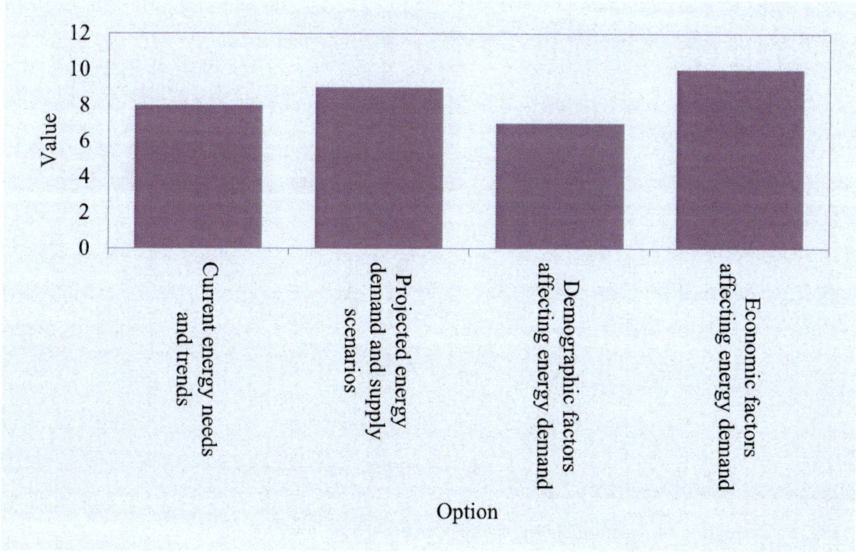

Fig. 6 Strategic options and decision-making weighting in energy policy

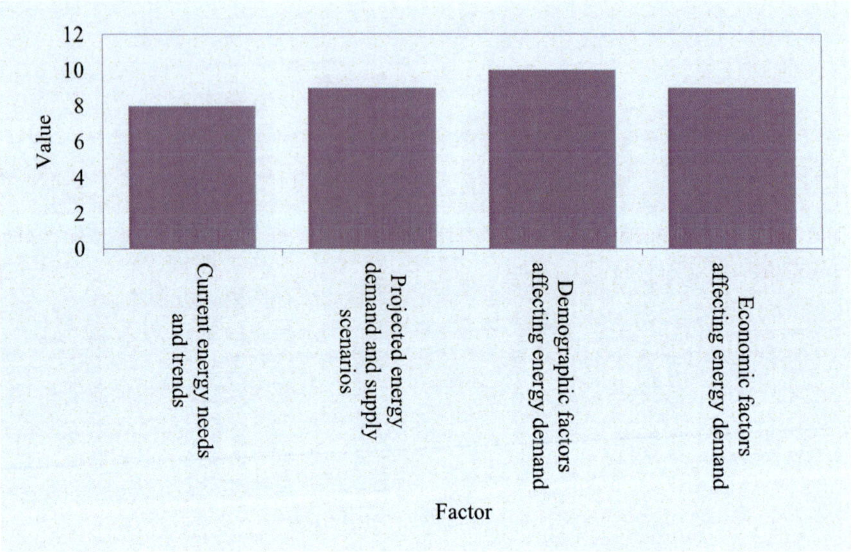

Fig. 7 Monitoring and effectiveness metrics in energy policy implementation

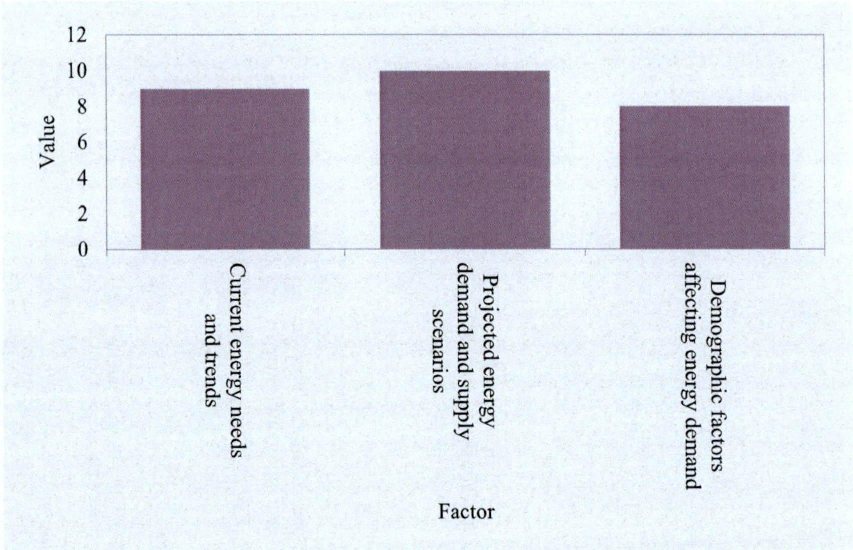

Fig. 8 Evaluation of stakeholder engagement in energy policy process

values is crucial for maintaining a realistic and effective energy policy framework. It ensures that policymakers prioritize resources and attention to the most pressing areas and are able to adapt to new challenges and opportunities as they arise. This dynamic approach to policy assessment and planning helps create responsive strategies that can better meet the goals of energy efficiency, sustainability, and security.

4 Energy and Carbon Supply Chain

The energy sector is a complex endeavor of activities and services that form the backbone of modern economies. It is a value chain that begins with the discovery and extraction of energy resources and extends all the way to the delivery of energy to consumers [21]. This value chain encompasses several critical components, each with its distinct role and importance:

- Exploration and production: This is the starting point of the energy value chain, where companies engage in the discovery of oil and gas reserves, drilling, and the completion of wells to bring raw energy resources to the surface [22].
- Refining: Raw crude oil is transformed into various usable products like gasoline and diesel in refineries. This process is vital for creating specific energy products that power vehicles, machinery, and many industrial processes [23].
- Transportation and logistics: Once extracted and refined, energy resources must be transported to various markets. This involves a network of pipelines, tankers, and trucks, each chosen for their cost efficiency and suitability for different types of energy products [24].
- Marketing and distribution: This stage involves the sales strategies and distribution networks that ensure energy products reach end consumers, whether through retail outlets or direct sales to large consumers like industries and utilities [25].
- Generation: Electricity generation is a diverse field within the energy sector, encompassing traditional coal and natural gas plants, nuclear power, and increasing contributions from renewable sources such as hydroelectric, wind, and solar [23].
- Transmission and distribution: This is the infrastructure that carries electricity from power plants to consumers. It includes the high-voltage transmission lines that carry electricity over long distances and the local distribution networks that deliver it to homes and businesses [26].
- Trading and marketing: Energy trading involves the buying and selling of energy commodities and derivatives, often to balance supply and demand across different regions, and can influence the price of energy on both wholesale and retail markets [27].
- Energy efficiency: This aspect focuses on reducing energy consumption and increasing efficiency across all stages of the energy value chain, from production to end use, through advanced technologies and policies [28].
- Renewable energy: With an increasing focus on sustainability, renewable energy development is a rapidly growing field. It includes technologies that harness wind, solar, and geothermal energy to produce electricity with a minimal environmental footprint [29].
- Energy storage: Essential for balancing supply and demand, energy storage technologies like batteries and pumped hydropower allow excess energy to be stored and then used when the demand is higher or generation is lower [30].

– Energy management and monitoring: Advanced systems and policies are employed to ensure that energy production and consumption are monitored and managed effectively, optimizing performance and reducing waste [31].
– Energy policy and regulation: The framework of policies and regulations governs the operation of the energy sector, including incentives for renewable energy, efficiency standards, and measures to control emissions [32].

The carbon emission value chain is a complex and interlinked sequence of activities contributing to the global carbon footprint, starting with energy generation [33]. Electricity production, particularly from coal and natural gas, is a major source of carbon emissions. Transportation, another significant emitter, encompasses vehicles, trains, planes, and ships that largely depend on fossil fuels. The industrial sector is not far behind, with manufacturing, refining, and petrochemical processes consuming vast amounts of energy, thus contributing heavily to emissions. Agricultural activities also add to the emission levels through practices involving livestock, fertilization, and the operation of machinery [34]. In the residential and commercial domains, carbon emissions result from heating, cooling, and lighting needs. Land use changes, such as deforestation and biofuel cultivation, lead to the release of carbon previously sequestered in forests and soil, exacerbating the issue. Additionally, waste management, particularly landfill operations and sewage treatment, generates methane, a potent greenhouse gas.

Conversely, the value chain also integrates vital emission reduction measures. Energy efficiency is achieved through the adoption of technologies and practices that curb energy use. Renewable energy sources like wind, solar, and geothermal offer greener alternatives to traditional fossil fuels, helping to reduce the overall emissions. The carbon capture and storage (CCS) technology is directed at capturing emissions directly from their sources and storing them to avoid atmospheric release [33]. Carbon offsetting complements this by funding emission reduction projects in other sectors to balance out emissions elsewhere [14]. Policy and regulation play a pivotal role in shaping the emission landscape, with governments implementing measures such as emissions trading schemes and carbon taxes to encourage reductions. Reforestation and afforestation efforts contribute to carbon sequestration, promoting biodiversity and climatic stability. Finally, adapting to climate change and enhancing resilience involves altering practices and bolstering infrastructure to cope with the changing climate, safeguarding ecosystems, communities, and built environments [35].

The comprehensive understanding and effective management of the carbon emission value chain are essential to global endeavors aimed at reducing greenhouse gas emissions and confronting climate change, a cornerstone for a sustainable and resilient future.

The key elements in the context of energy management, carbon emission reduction, and the integration of artificial intelligence (AI) across various sectors represent the overarching effort to create a more sustainable, efficient, and technologically advanced society, particularly in the context of energy and environmental challenges [19, 36, 37]. It encapsulates a comprehensive range of strategies, technologies, and policies aimed at enhancing efficiency, reducing environmental impact, and leveraging AI for better decision-making and operational effectiveness in these critical areas as follows [19, 36–42]:

- Predictive maintenance: Using AI to anticipate equipment failures and optimize maintenance schedules.
- Smart grid: Implementing AI-driven systems for efficient energy distribution and consumption management.
- Energy efficiency: Employing technologies, processes, and policies to reduce energy consumption and enhance efficiency.
- Renewable energy optimization: Maximizing the potential of renewable energy sources using AI algorithms.
- Carbon capture and storage (CCS): Capturing and storing carbon dioxide emissions from power plants and industrial processes.
- Climate policy and regulation: Developing policies and regulations to reduce emissions and mitigate climate change impacts.
- Carbon trading and offsetting: Reducing emissions in one sector by financing emission reductions in another sector.
- Agricultural emissions: Managing emissions from livestock, fertilizer use, and energy consumption in agricultural processes.
- Waste management: Addressing emissions from waste disposal in landfills and the treatment of sewage and other organic waste.
- Environmental impact assessment: Evaluating the potential environmental effects of projects and policies.
- Energy generation: Producing electricity from diverse energy sources.
- Industry, transportation, residential, and commercial: Reducing emissions across various sectors.
- Resource efficiency: Optimizing the use of natural resources in production processes.
- Environmental goals: Aligning policymaking with sustainability objectives.
- Data analytics, machine learning, predictive modeling, computer vision, natural language processing, and deep learning: Utilizing AI technologies for improved decision-making and efficiency.
- Internet of things (IoT), sensor networks, and cloud computing: Leveraging connected devices and cloud infrastructure for real-time data collection and processing.
- Robotics and autonomous systems: Deploying automated solutions for enhanced productivity and reduced emissions.
- Blockchain and distributed ledger technology: Using decentralized systems for secure data storage and sharing.
- Cybersecurity and privacy: Ensuring the protection of sensitive information in AI-driven systems.
- Ethical considerations: Evaluating the social and ethical implications of AI technologies in the energy and climate domains.

Table 2 Progressive hierarchy of sustainable technologies and environmental policies [20, 43, 44]

Hierarchy 1	Hierarchy 2	Hierarchy 3	Hierarchy 4	Hierarchy 5
Predictive maintenance	Smart grid	Energy efficiency	Renewable energy optimization	Carbon capture and storage (CCS)
Climate policy and regulation	Carbon trading and offsetting	Agricultural emissions	Waste management	Environmental impact assessment
Energy generation	Industry	Transportation	Residential and commercial	Resource efficiency
Environmental goals	Policymaking	Data analytics	Machine learning	Predictive modeling
Computer vision	Natural language processing	Deep learning	Internet of things (IoT)	Sensor networks
Cloud computing	Robotics	Autonomous systems	Blockchain	Distributed ledger technology
Cybersecurity	Privacy	Ethical considerations		

- Demand response management: Using AI to balance electricity supply and demand dynamically.
- Green building technologies: Incorporating sustainable design and construction methods to reduce environmental impact.
- Sustainable supply chain management: Optimizing supply chain operations for minimal environmental footprint.
- E-mobility and electric vehicles: Promoting the adoption of electric vehicles to reduce transportation-related emissions.

A range of concepts and technologies are organized into a five-tiered hierarchy (Table 2), starting from foundational elements like predictive maintenance, climate policy, and cybersecurity and progressively building up to more specific applications and broader impacts [20]. The second hierarchy introduces applications such as smart grid and robotics, which lead to more focused areas like energy efficiency and machine learning in the third tier [43]. This is further specialized in the fourth tier with concepts like renewable energy optimization and IoT. The final tier encapsulates outcomes or advanced applications, including carbon capture and storage (CCS), environmental impact assessment, and sensor networks, illustrating a clear progression from basic concepts to their nuanced applications and impacts [18].

Table 3 offers a comprehensive overview of different technological and environmental initiatives, categorized under various analytical aspects such as cost–benefit Analysis, environmental impact assessment, technical feasibility analysis, and stakeholder analysis [45]. Each row in Table 3 represents a different initiative or technology, illustrating how it fits into these analytical categories and its subcategories and target outcomes. For energy efficiency under cost–benefit analysis, the initiative is linked to Renewable Energy Optimization. This suggests an evaluation of the economic benefits versus the costs of optimizing renewable energy sources. For environmental impact assessment, it focuses on smart grid technologies, indicating an assessment of how the smart grid technology can impact the environment positively [46]. In technical feasibility analysis, the emphasis is on predictive

Table 3 A systematic value chain representation, including building blocks across the energy sector, carbon emission reduction, and AI integration [12, 28, 47, 54–56]

Cost–benefit analysis	Environmental impact assessment	Technical feasibility analysis	Stakeholder analysis	Subcategories	Target categories
Energy efficiency	Renewable energy optimization	Smart grid	Predictive maintenance	Energy generation, industry, transportation	Reduced energy consumption
Carbon capture and storage (CCS)	Climate policy and regulation	Carbon trading and offsetting	Agricultural emissions	Waste management, resource efficiency	Emissions reduction
Data analytics	Machine learning	Predictive modeling	Computer vision	Natural language processing, deep learning	Improved decision-making
Internet of things (IoT)	Sensor networks	Cloud computing	Robotics	Autonomous systems, blockchain, distributed ledger technology	Enhanced connectivity and automation
Cybersecurity	Privacy	Ethical considerations	Environmental goals	Policymaking, stakeholder engagement	Secure and responsible AI use

maintenance, assessing the technical viability of using predictive methods to maintain energy systems efficiently. Stakeholder analysis points to energy generation, industry, and transportation as the key sectors involved [47]. The subcategories include aspects like energy generation, and the ultimate target is reduced energy consumption. Carbon capture and storage's (CCS) [33] cost–benefit aspect is aligned with climate policy and regulation, evaluating the financial implications of climate policies on CCS. The environmental impact is examined through carbon trading and offsetting, assessing how these mechanisms affect the environment. Technical feasibility looks at agricultural emissions, focusing on the technical aspects of reducing emissions from agriculture. The stakeholders here include waste management and resource efficiency sectors. The target category is emissions reduction, underlining the primary goal of CCS.

Data analytics links to machine learning in cost–benefit analysis, evaluating the benefits and costs of implementing machine learning techniques in data analysis [48]. Environmental impact is assessed through predictive modeling, especially in terms of how these models can predict environmental outcomes [49]. Computer vision is highlighted in technical feasibility, exploring the technical potential of using computer vision in data analysis. Stakeholders involved are those in fields like natural language processing and deep learning [50]. The ultimate goal is improved decision-making. The Internet of Things (IoT) connects to sensor networks in the cost–benefit analysis, suggesting a focus on the economic evaluation of integrating sensor networks in IoT. Cloud computing is the focal point in environmental impact, assessing the environmental implications of using cloud technologies in IoT [51]. Robotics is considered in technical feasibility, examining the feasibility of integrating robotics into IoT systems [52]. Autonomous systems, blockchain, and distributed ledger technology are the key stakeholders, with the target being enhanced connectivity and automation. Finally, cybersecurity is discussed in the context of privacy under cost–benefit analysis, evaluating the trade-offs between cybersecurity measures and privacy concerns [42]. Ethical considerations are assessed in environmental impact, reflecting on the ethical implications of cybersecurity initiatives. Technical feasibility analysis focuses on environmental goals, exploring the feasibility of aligning cybersecurity with environmental objectives [28, 53]. Policymaking and stakeholder engagement are the critical stakeholders, and the target category is secure and responsible AI use.

Table 3 is categorized into various modern technological and environmental initiatives, providing a structured way to analyze them from multiple perspectives. This approach helps in understanding the multifaceted implications of these initiatives, considering their economic, environmental, technical, and social dimensions.

Table 4 is presented to discuss the relationship between various aspects of the value chain: cost–benefit analysis, environmental impact assessment, technical feasibility analysis, and stakeholder analysis. The subcategories further break down the elements within each aspect, while the target categories describe the overall goals to be achieved.

Table 4 Traceability system context diagram for the items in the value chain [12, 42, 47, 55–59]

System element	Inputs	Processing activities	Outputs	Stakeholders
Energy efficiency	Energy demand data, efficiency benchmarks	Analyze energy consumption, identify efficiency gaps	Efficiency improvement recommendations	Energy consumers, policymakers, utility companies
Emission reduction	Emission data, reduction targets	Calculate emissions, identify reduction opportunities	Emission reduction strategies	Industries, policymakers, environmental agencies
Renewable energy	Renewable energy generation data, forecasts	Analyze renewable energy potential, optimization	Renewable energy expansion plans	Energy producers, investors, regulators
AI in decision-making	Data from various sources, AI models, algorithms	Data analysis, predictive modeling, ML application	Data-driven insights, forecasts, decision support	Policymakers, business leaders, analysts
Connectivity and automation	IoT data, sensor networks, automation technologies	Integration, data processing, automation implementation	Enhanced connectivity, automated systems	Tech companies, end users, industries
Security and responsibility	Cybersecurity measures, privacy policies	Security assessments, ethical considerations	Secure systems, responsible AI use	IT professionals, businesses, end users

Energy efficiency focuses on optimizing energy usage [57]. It starts with inputs like energy demand data and efficiency benchmarks, which are analyzed to identify gaps in energy consumption. The processing activity involves scrutinizing these data to pinpoint areas where energy efficiency can be improved. The output of this process is a set of efficiency improvement recommendations [58]. The stakeholders involved include energy consumers, policymakers, and utility companies, all playing a role in implementing these recommendations and benefiting from enhanced energy efficiency.

For emission reduction, the inputs are emission data and reduction targets. The primary activity is calculating emissions and identifying opportunities to reduce them. This leads to the output of emission reduction strategies, which are essential for industries, policymakers, and environmental agencies, the key stakeholders. These strategies are critical for meeting regulatory requirements and achieving sustainability goals [21]. Renewable energy deals with the optimization of renewable energy sources. The inputs are renewable energy generation data and forecasts, which are analyzed to assess the potential and optimization of renewable energy sources. The output is a plan for the expansion of renewable energy, involving stakeholders such as energy producers, investors, and regulators. These plans are crucial for transitioning to cleaner energy sources and meeting environmental targets [59].

AI in decision-making underscores the role of AI in enhancing decision-making processes [17]. Various data sources, along with AI models and algorithms, serve as inputs. The processing activities involve data analysis, predictive modeling, and the application of machine learning. The output is data-driven insights, forecasts, and decision support tools, which are invaluable to policymakers, business leaders, and analysts [60]. These outputs assist in making more informed, data-backed decisions. Connectivity and automation focuses on the technological aspects of connectivity and automation; this element uses IoT data, sensor networks, and automation technologies as inputs. The processing activities include the integration and processing of these data, leading to the implementation of automated systems [61]. The output is enhanced connectivity and more sophisticated automated systems. The stakeholders here are tech companies, end-users, and various industries that depend on these technologies for operational efficiency. Security and responsibility addresses cybersecurity and ethical considerations. Inputs include cybersecurity measures and privacy policies [42]. The processing activities involve conducting security assessments and considering ethical implications. The output is the creation of secure systems and the responsible use of AI, benefiting stakeholders such as IT professionals, businesses, and end users. This element is critical in ensuring trust and safety in digital environments.

Table 5 [43, 47, 54, 55] provides an overview of the system context for each element. It outlines the inputs, processing activities, outputs, and stakeholders involved in each aspect of the energy sector, carbon emission reduction, and AI integration value chain. Energy policies, market conditions, and technology significantly impact energy efficiency. The integration of behavioral economics into energy efficiency policy is emerging, highlighting the importance of considering human behavior alongside physical, technical, and economic models [43]. The Paris

Table 5 Items in the energy sector, carbon emission reduction, and AI integration value chain [43, 47, 54, 55]

Element	Contextual factors	Actors and stakeholders	Interactions and dependencies
Energy efficiency	Energy policies, market conditions, technology	Energy consumers, policymakers, utility companies	Energy usage, efficiency initiatives, regulations
Emission reduction	Emission targets, climate policies, regulations	Industries, policymakers, environmental agencies	Emission monitoring, reduction strategies, policy
Renewable energy	Resource availability, policy incentives, costs	Energy producers, investors, regulators	Energy generation, investments, policy compliance
AI in decision-making	Data availability, AI capabilities, implementation	Policymakers, business leaders, analysts	Data analysis, predictive modeling, decision support
Connectivity and automation	IoT adoption, infrastructure, technology	Tech companies, end-users, industries	IoT data, sensor networks, automation technologies
Security and responsibility	Cybersecurity threats, data privacy concerns	IT professionals, businesses, end-users	Cybersecurity measures, privacy policies, ethics

Agreement also underscores the need for swift energy consumption reduction, with energy efficiency recognized as a crucial "win-win" solution worldwide. Energy consumers, policymakers, and utility companies are the key stakeholders in energy efficiency. Their decisions and actions greatly influence the implementation and success of energy efficiency initiatives. Energy usage, efficiency initiatives, and regulations are interdependent. The gap between the potential and actual adoption of energy efficiency measures is a significant issue in policy debates, requiring a multidisciplinary approach to understand and address it [54].

Emission targets, climate policies, and regulations are central to emissions reduction efforts. Federal climate policy, for instance, includes policies aimed at reducing greenhouse gases in the atmosphere [47]. Industries, policymakers, and environmental agencies play a crucial role in formulating and implementing emissions reduction strategies. Emission monitoring, reduction strategies, and policy implementation are closely linked. Different policy approaches, such as carbon pricing, technology subsidies, and performance standards, are considered to reduce emissions [47]. Factors like resource availability, policy incentives, and costs influence the adoption and implementation of renewable energy. State-level policies like renewable portfolio standards play a key role in governing the use of renewable energy sources. Energy producers, investors, and regulators are key stakeholders in renewable energy [55]. Their participation in decision-making processes is crucial for effective policy implementation. Energy generation, investments, and policy compliance are interconnected in the realm of renewable energy. The participatory processes in decision-making influence the implementation of renewable energy policies **[56].**

The availability of data, AI capabilities, and implementation approaches significantly impact the effectiveness of AI in decision-making. AI in data analysis involves algorithms that can recognize patterns in complex datasets and predict future outcomes [44]. Policymakers, business leaders, and analysts are key actors in leveraging AI for decision-making [42]. Data analysis, predictive modeling, and decision support are critical components of AI in decision-making. AI tools in data analytics provide accuracy, speed, and enhanced visualization for improved decision-making [62]. IoT adoption, infrastructure, and technology are fundamental to connectivity and automation. The growth of IoT and its dependence on strong connectivity are transforming various sectors [63]. Tech companies, end users, and industries are involved in the development and use of IoT and automation technologies. IoT data, sensor networks, and automation technologies are interdependent, shaping the future of connectivity and IoT applications [64].

Table 5 provides a context diagram for each element in the value chain. It outlines the contextual factors, actors and stakeholders, and interactions and dependencies involved in each aspect of the energy sector, carbon emission reduction, and AI integration.

5 Discussion

The integration of backcasting and AI in energy policy presents unique opportunities and challenges. The backcasting approach allows policymakers to envision a sustainable future, particularly focusing on renewable energy and reduced emissions, and work backward to identify practical steps toward these goals. This approach, while visionary, requires careful consideration of current technological and political realities to ensure feasibility. AI's role in energy policy is pivotal, offering predictive maintenance, smart grid management, and renewable energy optimization. However, the integration of AI also brings challenges, such as the need for substantial investments in technology, potential cybersecurity threats, and ethical concerns related to data privacy and AI decision-making.

Looking forward, the energy sector must focus on harnessing AI for renewable energy optimization, carbon capture, and effective policy implementation. Embracing a multidisciplinary approach is crucial, involving stakeholders from various sectors, including technology, policy, and environmental science. Future research should also explore the ethical implications of AI in energy policy and ways to mitigate potential risks.

6 Conclusion

This chapter has explored the use of backcasting and AI in developing effective and sustainable energy policies. The study underscores the importance of envisioning a sustainable future and methodically working backward to identify necessary policies and actions. The ASHES framework emerges as a robust tool for incorporating AI into energy policy, ensuring a comprehensive approach to address energy challenges. Future energy policies must adapt to rapidly evolving technological landscapes and embrace AI's potential while addressing associated risks and ethical concerns. The ongoing collaboration between policymakers, technologists, and environmentalists will be crucial in shaping a sustainable, efficient, and AI-integrated energy future. As the world continues to confront climate change and sustainability challenges, the insights from this study provide a valuable roadmap for future energy policy development, emphasizing innovation, collaboration, and a commitment to sustainable goals.

References

1. Bibri, S.E.: Backcasting in futures studies: a synthesized scholarly and planning approach to strategic smart sustainable city development. Eur. J. Futures Res. **6**, 13 (2018). https://doi.org/10.1186/s40309-018-0142-z
2. Vergragt, P.J., Quist, J.: Backcasting for sustainability: introduction to the special issue. Technol. Forecast. Soc. Chang. **78**, 747–755 (2011). https://doi.org/10.1016/j.techfore.2011.03.010
3. Kishita, Y., McLellan, B.C., Giurco, D., Aoki, K., Yoshizawa, G., Handoh, I.C.: Designing backcasting scenarios for resilient energy futures. Technol. Forecast. Soc. Chang. **124**, 114–125 (2017). https://doi.org/10.1016/j.techfore.2017.02.001
4. Sisto, R., Sica, E., Cappelletti, G.M.: Drafting the strategy for sustainability in universities: a Backcasting approach. Sustain. For. **12**, 4288 (2020). https://doi.org/10.3390/su12104288
5. Edgar, G., Kharazmi, O.A.: Systems evaluation of university-industry collaboration efficiency in Iran: current situation and proposed policy framework. J. Knowl. Econ. **14**, 645–675 (2023). https://doi.org/10.1007/s13132-021-00873-z
6. Denning, S.: Generating market-creating innovation by working backwards from the future. Strateg. Leadersh. **49**, 9–14 (2021). https://doi.org/10.1108/SL-10-2021-0094
7. Santos Coelho, R., Coelho, P.S., Antunes, P., Ramos, T.B.: Stakeholders perspectives on the use of indicators in water resources planning and related strategic environmental assessment. J. Env. Assmt. Pol. Mgmt. **21**, 1950001 (2019). https://doi.org/10.1142/S1464333219500017
8. Bibri, S.E.: A methodological framework for futures studies: integrating normative backcasting approaches and descriptive case study design for strategic data-driven smart sustainable city planning. Energy Inform. **3**, 31 (2020). https://doi.org/10.1186/s42162-020-00133-5
9. Shang, X., Chang, J., Yang, J., Ke, X., Duan, Z.: Life cycle sustainable assessment of natural vs artificial lightweight aggregates. J. Clean. Prod. **367**, 133064 (2022). https://doi.org/10.1016/j.jclepro.2022.133064
10. Letcher, M.R., Britton, J.: PV, or not PV: using backcasting to explore policy, market and governance implications of local decarbonisation pathways such as urban PV. Renew. Energy Focus. **44**, 244–258 (2023). https://doi.org/10.1016/j.ref.2022.12.004
11. Quist, J., Vergragt, P.: Past and future of backcasting: the shift to stakeholder participation and a proposal for a methodological framework. Futures. **38**, 1027–1045 (2006). https://doi.org/10.1016/j.futures.2006.02.010
12. Qudrat-Ullah, H.: A review and analysis of green energy and the environmental policies in South Asia. Energies. **16**, 7486 (2023). https://doi.org/10.3390/en16227486
13. Revell, P., Dinnie, E.: Community resilience and narratives of community empowerment in Scotland. Community Dev. J. **55**, 218–236 (2020). https://doi.org/10.1093/cdj/bsy038

14. Kumar, R., Ojha, K., Ahmadi, M.H., Raj, R., Aliehyaei, M., Ahmadi, A., Nabipour, N.: A review status on alternative arrangements of power generation energy resources and reserve in India. Int. J. Low-Carbon Technol. **15**, 224–240 (2020). https://doi.org/10.1093/ijlct/ctz066
15. Zhukovskiy, Y., Tsvetkov, P., Buldysko, A., Malkova, Y., Stoianova, A., Koshenkova, A.: Scenario modeling of sustainable development of energy supply in the Arctic. Resources. **10**, 124 (2021). https://doi.org/10.3390/resources10120124
16. Morgan, E.A., Nalau, J., Mackey, B.: Assessing the alignment of national-level adaptation plans to the Paris agreement. Environ. Sci. Pol. **93**, 208–220 (2019). https://doi.org/10.1016/j.envsci.2018.10.012
17. Döme, V.: A global-scale study on decision making in renewable energy policy: internal and external factors driving the adoption of feed-in tariffs and renewable portfolio standards. Environ. Policy Gov. **1**, 1–15 (2023). https://doi.org/10.1002/eet.2085
18. Danish, M.S.S., Senjyu, T.: Green building efficiency and sustainability indicators. In: Solanki, A., Nayyar, A. (eds.) Green Building Management and Smart Automation, pp. 128–145. IGI Global, Hershey (2020). https://doi.org/10.4018/978-1-5225-9754-4
19. Danish, M.S.S., Elsayed, M.E.L., Ahmadi, M., Senjyu, T., Karimy, H., Zaheb, H.: A strategic-integrated approach for sustainable energy deployment. Energy Rep. **6**, 40–44 (2020). https://doi.org/10.1016/j.egyr.2019.11.039
20. Danish, M.S.S., Senjyu, T.: AI-Enabled Energy Policy for a Sustainable Future, https://www.researchgate.net/publication/369850945_AI-Enabled_Energy_Policy_for_a_Sustainable_Future (2023)
21. Adalı, Z., Danish, M.S.S.: Investigation of the nexus between the electricity consumption and the ecological footprint. In: Dinçer, H., Yüksel, S. (eds.) Circular Economy and the Energy Market: Achieving Sustainable Economic Development through Energy Policy, pp. 79–89. Springer International Publishing, Cham (2022). https://doi.org/10.1007/978-3-031-13146-2_7
22. Danish, M.S.S., Senjyu, T., Sabory, N.R., Danish, S.M.S., Ludin, G.A., Noorzad, A.S., Yona, A.: Afghanistan's aspirations for energy independence: water resources and hydropower energy. Renew. Energy. **113**, 1276–1287 (2017). https://doi.org/10.1016/j.renene.2017.06.090
23. Chow, J., Kopp, R.J., Portney, P.R.: Energy resources and global development. Science. **302**, 1528–1531 (2003). https://doi.org/10.1126/science.1091939
24. Ngo, C., Natowitz, J.: Our Energy Future: Resources, Alternatives and the Environment. Wiley, Hoboken (2016)
25. Sokolinskaya, N., Danish, M.S.S.: Development Russian financial markets: evidence from energy companies from 1990 to 2020. In: Yüksel, S., Dinçer, H. (eds.) Strategic Approaches to Energy Management: Current Trends in Energy Economics and Green Investment, pp. 39–50. Springer International Publishing, Cham (2021). https://doi.org/10.1007/978-3-030-76783-9_4
26. Furukakoi, M., Danish, M.S.S., Howlader, A.M., Senjyu, T.: Voltage stability improvement of transmission systems using a novel shunt capacitor control. Int. J. Emerg. Electr. Power Syst. **19**, 1–12 (2018). https://doi.org/10.1515/ijeeps-2017-0112
27. Danish, M.S.S., Senjyu, T.: Shaping the future of sustainable energy through AI-enabled circular economy policies. Circ. Econ. **2**, 100040 (2023). https://doi.org/10.1016/j.cec.2023.100040
28. Danish, M.S.S., Esteves, E.G.Z.: Human resource management to improve the quality of energy efficiency of the enterprise. In: Dincer, H., Yüksel, S. (eds.) Economic Development and the Environmental Ecosystem: the Role of Energy Policy in Economic Growth, pp. 141–151. Springer Nature, Cham (2023). https://doi.org/10.1007/978-3-031-26596-9_12
29. Bhuiyan, M.A.A., An, J., Mikhaylov, A., Moiseev, N., Danish, M.S.S.: Renewable energy deployment and COVID-19 measures for sustainable development. Sustain. For. **13**, 4418 (2021). https://doi.org/10.3390/su13084418
30. Ahmadi, M., Lotfy, M.E., Danish, M.S.S., Ryuto, S., Yona, A., Senjyu, T.: Optimal multi-configuration and allocation of SVR, capacitor, centralised wind farm, and energy storage system: a multi-objective approach in a real distribution network. IET Renew. Power Gener. **13**, 762–773 (2019). https://doi.org/10.1049/iet-rpg.2018.5057

31. Danish, M.S.S., Senjyu, T., Funabashi, T., Ahmadi, M., Ibrahimi, A.M., Ohta, R., Rashid Howlader, H.O., Zaheb, H., Sabory, N.R., Sediqi, M.M.: A sustainable microgrid: a sustainability and management-oriented approach. Energy Procedia. **159**, 160–167 (2019). https://doi.org/10.1016/j.egypro.2018.12.045
32. Danish, M.S.S., Ibrahimi, A.M., Ahmadi, M., Shams, S., Fedayi, H., Senjyu, T.S.: Energy strategy and policy development process. In: Eco-Friendly and Agile Energy Strategies and Policy Development, pp. 1–27. IGI Global, Hershey (2022). https://doi.org/10.4018/978-1-7998-9502-2.ch001
33. Cao, K., Xu, X., Wu, Q., Zhang, Q.: Optimal production and carbon emission reduction level under cap-and-trade and low carbon subsidy policies. J. Clean. Prod. **167**, 505–513 (2017). https://doi.org/10.1016/j.jclepro.2017.07.251
34. Kamal, A., Al-Ghamdi, S.G., Koç, M.: Role of energy efficiency policies on energy consumption and CO2 emissions for building stock in Qatar. J. Clean. Prod. **235**, 1409–1424 (2019). https://doi.org/10.1016/j.jclepro.2019.06.296
35. Danish, M.S.S., Senjyu, T.S. (eds.): System of green resilience eco-oriented land uses in urban socio-ecosystems. In: Eco-Friendly Energy Processes and Technologies for Achieving Sustainable Development, pp. 1–23. IGI Global, Hershey (2021). https://doi.org/10.4018/978-1-7998-4915-5
36. Danish, M.S.S., Sabory, N.R., Danish, S.M.S., Ludin, G.A., Yona, A., Senjyu, T.: An open-door immature policy for rural electrification: a case study of Afghanistan. Int. J. Sustain. Green Energy. **6**(8–13) (2016). https://doi.org/10.11648/j.ijrse.s.2017060301.12
37. Khare, V.R., Danish, M.S.S., Khosravy, M., Matin, A., Mikhaylov, A., Senjyu, T.: Energy efficiencies model for thermal comfort in urban applications. J. Environ. Sci. Revolut. **1–17** (2022). https://doi.org/10.37357/1068/jesr/3.1.01
38. Bhuiyan, M.A., Dinçer, H., Yüksel, S., Mikhaylov, A., Danish, M.S.S., Pinter, G., Uyeh, D.D., Stepanova, D.: Economic indicators and bioenergy supply in developed economies: QROF-DEMATEL and random forest models. Energy Rep. **8**, 561–570 (2022). https://doi.org/10.1016/j.egyr.2021.11.278
39. Danish, M.S.S., Senjyu, T., Ibrahimi, A.M., Ahmadi, M., Howlader, A.M.: A managed framework for energy-efficient building. J. Build. Eng. **21**, 120–128 (2019). https://doi.org/10.1016/j.jobe.2018.10.013
40. Danish, M.S.S., Senjyu, T., Zaheb, H., Sabory, N.R., Ibrahimi, A.M., Matayoshi, H.: A novel transdisciplinary paradigm for municipal solid waste to energy. J. Clean. Prod. **233**, 880–892 (2019). https://doi.org/10.1016/j.jclepro.2019.05.402
41. Danish, M.S.S.: AI and expert insights for sustainable energy future. Energies. **16**, 3309 (2023). https://doi.org/10.3390/en16083309
42. Danish, M.S.S.: AI in energy: overcoming unforeseen obstacles. AI. **4**, 406–425 (2023). https://doi.org/10.3390/ai4020022
43. Della Valle, N., Bertoldi, P.: Promoting energy efficiency: barriers, societal needs and policies. Front. Energy Res. **9** (2022)
44. Danish, M.S.S.: A framework for modeling and optimization of data-driven energy systems using machine learning. IEEE Trans. Artif. Intell., 1–10 (2023). https://doi.org/10.1109/TAI.2023.3322395
45. Danish, M.S.S., Yona, A., Senjyu, T.: Pre-design and life cycle cost analysis of a hybrid power system for rural and remote communities in Afghanistan. J. Eng IET. **2014**, 438–444 (2014). https://doi.org/10.1049/joe.2014.0172
46. Lu, R., Hong, S.H.: Incentive-based demand response for smart grid with reinforcement learning and deep neural network. Appl. Energy. **236**, 937–949 (2019). https://doi.org/10.1016/j.apenergy.2018.12.061
47. Rountree, V., Baldwin, E.: State-level renewable energy policy implementation: how and why do stakeholders participate? Front. Commun. **3** (2018)
48. Nazari, Z., Nazari, M., Danish, M.S.S., Kang, D.: Evaluation of class noise impact on performance of machine learning algorithms. Int. J. Comput. Sci. Network Secur. **18**, 148–153 (2018)

49. Syed, M.A., Khalid, M.: An intelligent model predictive control strategy for stable solar-wind renewable power dispatch coupled with hydrogen Electrolyzer and battery energy storage. Int. J. Energy Res. **2023**, e4531054 (2023). https://doi.org/10.1155/2023/4531054
50. Bentsen, L.Ø., Warakagoda, N.D., Stenbro, R., Engelstad, P.: Probabilistic Wind Park power prediction using Bayesian deep learning and generative adversarial networks. J. Phys. Conf. Ser. **2362**, 012005 (2022). https://doi.org/10.1088/1742-6596/2362/1/012005
51. Feng, Y., Shi, W., Shi, W., Cheng, G., Huang, J., Liu, Z.: Benchmarking framework for command and control mission planning under uncertain environment. Soft. Comput. **24**, 2463–2478 (2020). https://doi.org/10.1007/s00500-018-03732-3
52. Riva, G., Wiederhold, B.K.: Human–robot confluence: toward a humane robotics. Cyberpsychol. Behav. Soc. Netw. **24**, 291–293 (2021). https://doi.org/10.1089/cyber.2021.29215.gri
53. Danish, S.M.S., Zaheb, H., Sabori, N.R., Karimy, H., Faiq, A.B., Fedayi, H., Senjyu, T.: The road ahead for municipal solid waste management in the 21st century: a novel-standardized simulated paradigm. In: IOP Conference Series: Earth and Environmental Science, p. 012009. IOP Publishing, Seoul (2019). https://doi.org/10.1088/1755-1315/291/1/012009
54. Vangala, S., Hung, K., South, D.: Revisiting the Biden Administration's approach to climate change. Climate Energy. **39**, 1–12 (2022). https://doi.org/10.1002/gas.22299
55. Bishoge, O.K., Kombe, G.G., Mvile, B.N.: Community participation in the renewable energy sector in Tanzania: community participation in the renewable energy sector. Int. J. Sustain. Energy Plan. Manag. **28**, 121–134 (2020). https://doi.org/10.5278/ijsepm.4477
56. Damayra, F., Khatib, T.: Assessment of innovation policy standards' impact on local development of renewable energy in Palestinian local government units. Renew. Energy. **187**, 177–192 (2022). https://doi.org/10.1016/j.renene.2022.01.065
57. Danish, M.S.S., Senjyu, T., Ahmadi, M., Ludin, G.A., Ahadi, M.H., Karimy, H., Khosravy, M.: A review on energy efficiency for pathetic environmental trends mitigation. J. Sustain. Outreach 2, 1–8 (2021). 10.37357/1068/jso.2.1.01
58. Danish, M.S.S., Senjyu, T.S.: Green building efficiency and sustainability indicators. In: Green Building Management and Smart Automation, pp. 128–145. IGI Global, Hershey (2020). https://doi.org/10.4018/978-1-5225-9754-4.ch006
59. Danish, M.S.S., Senjyu, T.: AI-enabled energy policy for a sustainable future. Sustain. For. **15**, 7643 (2023). https://doi.org/10.3390/su15097643
60. Danish, M.S.S., Nazari, Z., Senjyu, T.: AI-coherent data-driven forecasting model for a combined cycle power plant. Energy Convers. Manag. **286**, 117063 (2023). https://doi.org/10.1016/j.enconman.2023.117063
61. Danish, M.S.S., Senjyu, T., Nazari, M., Zaheb, H., Nassor, T.S., Danish, S.M.S., Karimy, H.: Smart and sustainable building appraisal. Journal of sustainable energy. Revolution. **2**(1–5) (2021). https://doi.org/10.37357/1068/jser.2.1.01
62. Emma: AI in Data Analytics: Transforming Decision-Making, https://www.pixelcrayons.com/blog/digital-transformation/ai-in-data-analytics-transforming-decision-making/. (Accessed 12 Mar 2023)
63. Nazim, S.F., Danish, M.S.S., Senjyu, T.: A brief review of the future of smart mobility using 5G and IoT. J. Sustain. Outreach. **19–30** (2022). https://doi.org/10.37357/1068/jso/3.1.02
64. Adli, H.K., Husin, K.A.K., Hanafiah, N.H.M., Remli, M.A., Ernawan, F., Wirawan, P.W.: Forecasting and analysis of solar power output from integrated solar energy and IoT system. In: 2021 5th International Conference on Informatics and Computational Sciences (ICICoS), pp. 222–226. IEEE, Semarang (2021). https://doi.org/10.1109/ICICoS53627.2021.9651831

A Blueprint for Sustainable Electrification by Designing and Implementing PV Systems in Small Scales

Hasan Dinçer, Abdul Matin Ibrahimi, Mikaeel Ahmadi, and Mir Sayed Shah Danish

1 Introduction

Solar energy has recently become a leading force in the shift toward renewable energy, with solar photovoltaic (PV) capacity experiencing rapid growth globally [1]. This growth is driven by significant improvements in the efficiency and affordability of solar panels. Key adopters include China, the United States, and India, supported by favorable government policies and an increasing focus on environmental sustainability [2]. The European Union is also actively increasing solar installations as part of its commitment to achieve climate neutrality by 2050 [3, 4]. The global solar energy market is expected to continue expanding, propelled by ongoing technological advancements and rising investments in renewable energy. When compared to other renewable sources like wind, hydroelectric, and biomass, solar energy presents distinct advantages and challenges. Solar PV systems, for instance, are advantageous due to their lack of mechanical moving parts, which translates to lower maintenance needs and longer lifespans. They are also highly scalable, from small residential setups to large-scale solar farms. However, the variability of solar energy and its dependency on weather conditions require advancements in energy storage to ensure a consistent energy supply. Hydroelectric power, while offering a steadier energy supply, faces geographical and environmental

H. Dinçer (✉)
School of Business, Istanbul Medipol University, Istanbul, Turkey
e-mail: hdincer@medipol.edu.tr

A. M. Ibrahimi · M. Ahmadi
Research Promotion Unit, Co-Creation Management Department, University of the Ryukyus, Okinawa, Japan

M. S. S. Danish
Energy Systems (Chubu Electric Power) Funded Research Division, Nagoya University, Nagoya, Japan

M. S. S. Danish (ed.), *Unified Vision for a Sustainable Future*,
https://doi.org/10.1007/978-3-031-53574-1_7

limitations [5]. Wind energy provides a more consistent resource than solar but has its challenges, including noise pollution, visual impact, and potential interference with bird migration [6]. Each renewable energy source plays a unique and vital role in creating a diverse, sustainable, and resilient energy mix for the future.

The integration of solar energy systems into the electricity market is a pivotal step toward addressing the dual challenges of energy security and climate change [7]. As a clean and renewable energy source, solar power offers a sustainable alternative to fossil fuels, significantly reducing greenhouse gas emissions and mitigating the impacts of climate change [8]. The scalability and decreasing costs of solar technology make it increasingly accessible and economically viable, encouraging its adoption in diverse settings, from residential rooftops to large-scale solar farms [9, 10]. In the electricity market, solar energy contributes to diversifying the energy mix, enhances grid resilience, and facilitates the transition to a low-carbon economy [11]. Moreover, the inherent intermittency of solar power drives innovations in energy storage and grid management [12], paving the way for a more flexible and reliable energy system capable of meeting the dynamic demands of modern society while preserving environmental integrity [13].

The literature focuses on the growing electricity demand and highlights the challenges of high energy costs, dangerous temperatures, and air pollution, emphasizing the need for grid reliability and public health protection [14]. It delves into the potential of distributed energy resources (DERs) to mitigate these impacts, including distributed generation, energy efficiency, demand response, and energy storage [1]. The review particularly notes the effectiveness of solar photovoltaic (PV) systems in reducing peak energy demand due to their low installation costs and peak generation during mid-day high-demand periods [15].

Reference [16] explores the complex dynamics of integrating DERs into the electrical system. It discusses the need for multi-objective optimization techniques for siting various types of DERs, considering the complexities introduced by small-scale generation resources such as PV systems. A review study touches on the importance of solar carve-outs in renewable portfolio standards, citing examples from US states like Arizona and Nevada, and notes the challenges in valuing the locational benefits of solar energy within these frameworks [17]. Furthermore, the study in [18] addresses the limitations of current systems in integrating solar and wind resources, underscoring the need for improved system flexibility. At the same time, another study emphasizes the importance of assessing the impact of solar PV on system flexibility requirements, particularly in light of the shift in diurnal peak load toward evening hours and the consequent challenges for dispatchable generators [19].

In 2021, Adli et al. [20] introduced a novel approach combining autoregressive integrated moving average (ARIMA) and artificial neural network (ANN) models. This method, leveraging a tracking system powered by the Internet of things (IoT) aimed to improve the energy production efficiency in solar power systems. The model was evaluated using the mean absolute percentage error (MAPE) metric, achieving an accuracy measure of 32.0%. A year earlier, in 2018, Qing and Niu [21] employed the long short-term memory (LSTM) network for day-ahead forecasting of sun radiation, utilizing meteorological information. This approach was particularly notable for its use of advanced recurrent neural networks in predicting solar radiation, with an accuracy measure, as indicated by the root mean square error (RMSE), of 18.34%. In 2017, Sivaneasan et al. [22] combined ANN and fuzzy logic for solar power forecasting. Their model integrated a triple-layer backpropagation (BP) network with fuzzy preprocessing, enhancing the ANN's predictive capabilities. The model's effectiveness was demonstrated through a MAPE of 29.6%, highlighting the synergy between fuzzy logic and neural networks in handling uncertain and complex solar forecasting scenarios.

Gutierrez-Corea et al. [23], in 2016, focused on forecasting global solar irradiance using an ANN model that considered various atmospheric and environmental factors. Their approach underlined the importance of comprehensive data inputs for accurate solar irradiance prediction, achieving a normalized RMSE (nRMSE) of 20%. In 2015, Abuella and Chowdhury [24] employed a feed-forward ANN (FF-ANN) with a novel approach involving preprocessing of input data, clustering, and the elimination of night hours to enhance prediction accuracy. Their method yielded a high R-squared value of 96.65%, showcasing the effectiveness of data preprocessing and clustering techniques in improving the accuracy of solar power predictions. In 2014, Dahmani et al. [25] used an ANN-multilayer perceptron (ANN-MLP) model to evaluate radiation data, including horizontal extraterrestrial irradiance, solar declination, and zenith angle. Their model, with an RMSE of 8.81%, demonstrated the use of incorporating comprehensive solar radiation data in enhancing forecasting accuracy. In 2013, Marquez et al. [26] developed a radiation forecasting model using an ANN. This model incorporated data from velocimetry, cloud indexing, and solar irradiation, presenting a versatile approach to solar radiation prediction. The model's RMSE ranged between 5% and 25%, illustrating the variability and potential challenges in solar radiation forecasting. These studies contribute to the evolving field of solar power forecasting [27], highlighting the importance of diverse data inputs [28], advanced computational models [29], and innovative processing techniques in improving the accuracy and efficiency of solar energy predictions [30].

This study is systematically structured into several sections. Section 2 explores the essential factors in planning and designing a PV system, including site selection, climatic considerations, and technical specifications. Section 3 presents a case study, illustrating the practical application of these principles in electrifying a rural community. In Sect. 4, calculations and estimating the balance of the PV system through the methodologies and calculations critical is discussed for the design and implementation of PV systems, covering aspects from solar irradiation to economic analysis. The discussion and lessons learned in Sect. 5 reflect on the insights and challenges encountered, providing a valuable context for understanding the study's practical implications. Finally, the study culminates in Sect. 6 with the conclusion, synthesizing the key findings and underscoring the significance of solar PV systems in addressing contemporary energy and environmental challenges.

2 Planning and Designing Photovoltaic (PV) System

The planning and design of a PV system involve a comprehensive analysis of various factors. Each factor plays a crucial role in determining the effectiveness, sustainability, and economic viability of the solar installation. Proper evaluation and integration of these factors ensure that the PV system not only meets the immediate energy needs but also contributes positively to the broader goals of energy sustainability and climate change mitigation.

- Site location and climatic conditions [31]: The geographical location of the PV system significantly influences its performance. Factors such as the intensity of sunlight, duration of solar exposure, and climatic conditions (like humidity, temperature, and seasonal variations) play a critical role. In areas with higher solar irradiance, solar panels will naturally produce more energy. However, extreme conditions like high temperatures can reduce the efficiency of the panels. Additionally, regions with frequent cloud cover or higher latitudes may require more panels or specifically designed systems to be as effective.
- Available roof or land space [32]: The physical space available for installing solar panels dictates the size and the layout of the PV system. Rooftop installations are common in urban settings, where space is limited. The roof's size, shape, and structural integrity are important considerations. In contrast, ground-mounted systems are more feasible in rural or industrial areas where land is more abundant. These systems can be more extensive and are easier to orient and angle for optimal sun exposure.
- Electrical load and energy consumption patterns [33]: Understanding the energy consumption patterns of the building or facility is crucial. This involves analyzing the electrical load, peak energy usage times, and total energy consumption over different seasons. These data help in designing a PV system that adequately meets the energy needs without significant wastage or shortfall.
- Local building codes and regulations [34]: Compliance with local building codes and regulations is nonnegotiable. These may include restrictions on the height and placement of solar panels, aesthetic considerations, and safety standards. Understanding and adhering to these regulations prevent legal complications and ensure the safety and reliability of the solar installation.
- Availability and cost of electrical grid connection [35]: The proximity to and the cost of connecting to the local electrical grid can influence the design of the PV system. In grid-connected systems, excess energy can be fed back into the grid, possibly earning credits or income. In more remote areas, off-grid systems with battery storage might be more viable, although these often come with higher initial costs.
- Availability of financial incentives and subsidies [36]: Many regions offer incentives, rebates, or subsidies to encourage solar energy adoption. These financial incentives can significantly reduce the initial investment and improve the return on investment of the solar system. Understanding what incentives are available and how to qualify for them is a critical part of the planning process.
- Orientation, tilt, and shading of the solar panels [37]: The orientation (direction) and tilt (angle) of solar panels directly affect their efficiency. Panels are typically

oriented to maximize exposure to the sun, with adjustments made based on the latitude of the location. Additionally, shading from nearby trees, buildings, or other structures can significantly reduce the efficiency of the panels. A site survey is essential to identify and mitigate potential shading issues.

- Availability and cost of battery storage system [38]: For off-grid systems or grid-tied systems with energy storage, the availability and cost of batteries are crucial. Batteries store excess energy generated during peak sun hours for use during low sun hours or at night. The type, capacity, and cost of batteries vary widely and must be matched with the system's energy production and consumption needs.
- Technical specifications of the solar panels and inverters [39]: The choice of solar panels and inverters is influenced by their efficiency, durability, warranty, and cost. Panels with higher efficiency might be more expensive but can generate more power in a smaller space. Similarly, inverters, which convert the DC power produced by the panels into AC power for home appliances, must be chosen based on their efficiency, compatibility with the panels, and the overall energy requirements of the system.
- Maintenance and operational requirements [40]: The long-term maintenance and operational needs of the PV system must be considered. This includes regular cleaning of panels, monitoring system performance, and replacing components like inverters or batteries over time. A low-maintenance and reliable system might have higher upfront costs but can be more cost-effective in the long run.
- Future expansion plans and scalability [41]: Planning for future expansion or scalability of the system is prudent. This might include preparing for increased energy needs, technological advancements, or the addition of energy storage. Designing a system that is easily scalable can provide more flexibility and adaptability in the future.
- System efficiency and energy generation estimates [42]: Estimating the system's efficiency and the total energy generation is vital for calculating the return on investment and ensuring that the system meets the energy needs. This involves considering the efficiency of the panels, the loss in energy conversion, and the potential impact of environmental factors.
- Environmental impact and sustainability considerations [43]: The environmental impact and sustainability of the PV system are increasingly important. These encompass the life cycle analysis of the solar panels, the recyclability of components, and the overall carbon footprint of the system. A sustainable system not only contributes to reducing greenhouse gas emissions but also aligns with broader environmental goals and regulations.

In this chapter, the scoring criteria for evaluating the factors involved in planning and designing a PV system are crucial in ensuring that the system is efficient, cost-effective, compliant with regulations, environmentally sustainable, and adaptable to various contexts. Each criterion carries a specific assumed weight, signifying its relative importance in the overall assessment of the PV system. The impact on system efficiency and performance (35%) criterion addresses how each factor influences the PV system's ability to convert sunlight into electrical energy effectively and reliably. Factors like site location, panel orientation, and technical specifications directly affect the amount of energy generated and the system's overall performance. Assigning the highest weight (35%) to this criterion reflects the primary purpose of a PV system – to generate electricity efficiently. A system's efficiency determines its economic viability and technological effectiveness, making this aspect paramount in the planning process.

The cost implications (30%) of the financial aspect encompass the initial capital expenditure, ongoing operational costs, potential savings, and the impact of financial incentives. The cost of components like solar panels, inverters, and battery storage systems, along with installation and maintenance expenses, significantly influences the economic feasibility of the PV system. Given that the adoption of solar energy often hinges on its cost competitiveness with traditional energy sources, it is justified that cost implications hold a substantial weight of 30%. This ensures a balance between economic feasibility and technical performance.

The regulatory and compliance importance (15%) with local building codes, electrical standards, and environmental regulations is nonnegotiable. This criterion ensures that the system meets all legal requirements, which is crucial for the system's legality and long-term operational stability. While regulatory compliance is essential, it is given a moderate weight (15%) since it is a baseline requirement rather than a performance enhancer. This weight acknowledges the necessity of compliance without overshadowing the system's efficiency and cost factors.

The environmental and sustainability impact (10%) evaluates the system's environmental footprint, including its life cycle impact and contributions to sustainability goals, like reducing greenhouse gas emissions. Solar energy systems are inherently aligned with environmental sustainability objectives. However, the weight of 10% for this criterion suggests that while environmental impact is important, it is one of several critical considerations. This weight reflects a balanced approach where environmental considerations are integral but not the sole focus.

The ease of implementation and flexibility (10%) consider the practical aspects of implementing the PV system, including the adaptability of the system to different contexts and the ease with which it can be installed and integrated into existing infrastructures. The adaptability and ease of implementation are given a weight of 10%, indicating their importance in the broader context of system design and deployment. This acknowledges that while technical, financial, and regulatory aspects are crucial, the practicality of implementing the system efficiently and flexibly also plays a significant role in its overall success.

The distribution of weights across these criteria in Fig. 1 and Table 1 reflects a balanced approach to evaluating the myriad factors that influence the planning and design of a PV system. This method ensures that all key aspects, from efficiency and cost to regulatory compliance, environmental impact, and practical implementation, are considered in concert, leading to the development of well-rounded, effective, and sustainable solar energy systems.

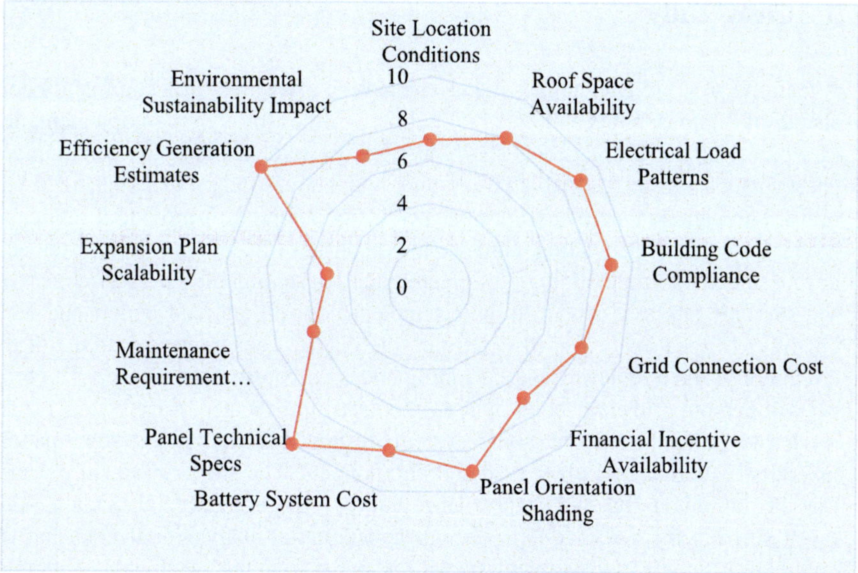

Fig. 1 Key factors for photovoltaic system planning and design

Table 1 Evaluating key factors influencing PV system efficiency and viability

Factor	Impact description
Site location conditions	High impact on system efficiency, moderate cost implications
Roof space availability	Critical for system installation, high-cost implications
Electrical load patterns	Critical for system sizing and efficiency
Building code compliance	Critical for compliance and legal aspects
Grid connection cost	High impact on cost and system design
Financial incentive availability	High impact on cost but variable importance based on location
Panel orientation shading	Critical for system performance and efficiency
Battery system cost	High impact on cost, system flexibility, and performance
Panel technical specs	Critical for system performance and longevity
Maintenance requirement operations	Moderate impact on long-term system performance and cost
Expansion plan scalability	Moderate impact on long-term viability and adaptability
Efficiency generation estimates	Critical for determining system effectiveness and ROI
Environmental sustainability impact	High importance for environmental sustainability but moderate direct impact on system performance

3 Case Study

A tentative case study examines the electrification of a small rural community consisting of 32 households and essential public facilities through the implementation of a photovoltaic (PV) system. The objectives are to understand the community's energy needs, design a suitable PV system, and analyze the project's financial viability. The significance of this project lies in its potential to provide a replicable model for rural electrification, enhance the quality of life, and promote sustainable development. The methodology for planning and designing the PV system involves a systematic site assessment to collect meteorological data and understand local conditions. Load analysis will determine the energy consumption patterns of the community, and a review of local building codes will ensure compliance. The project will explore grid connection options and determine the necessary system capacity based on collected data. The system design will encompass solar panels, inverters, batteries, and other essential components. The site assessment revealed that the community has sufficient solar radiation levels and minimal shading, making it suitable for a PV system. Energy consumption analysis indicated that the community's energy needs fluctuate, with peak usage in the evenings. Local building codes allow for the installation of PV systems, with certain restrictions to maintain the rural aesthetic.

Comparing the community's energy needs with the potential solar energy production, it was found that a PV system could effectively meet the demand. The analysis also indicated that battery storage would be necessary to ensure power availability during non-sunlight hours, and it would be cost-effective given the high costs associated with the grid connection. Based on the recommendations, it is recommended to install a PV system with a capacity that exceeds the community's average consumption by at least 20% to accommodate future growth. The system should include a robust battery storage solution to manage the evening peak loads. Financial analysis suggested that, with available subsidies, the project has a favorable payback period. Electrification of the rural community with a PV system is not only feasible but also economically and environmentally advantageous. The proposed system design and capacity will satisfy the current and future energy needs, contributing to the community's sustainability and resilience.

For conducting the process, supporting information such as detailed load analysis data, site assessment results, system design schematics, and financial analysis spreadsheets will be included. Such a project will require engagement with the community to ensure their needs are met and to foster local ownership of the PV system. Regular updates and transparent communication will be crucial for the project's success, along with training for local maintenance and monitoring of the system after installation.

An exhaustive roadmap for the planning, design, and implementation of a PV system, from project initiation to sign-off, is outlined based on the research studies discussed in Sect. 2 (Planning and Designing a Photovoltaic (PV) System) and the studied literature in this paper. This roadmap integrates insights from various

Table 2 PV system development and implementation, a step-by-step checklist

Step	Building blocks
Conduct a site assessment	Meteorological data, location, topography, shading analysis
Determine energy consumption	Load analysis, energy consumption patterns
Review local building codes and regulations	Building codes, electrical codes, fire codes
Assess grid connection options	Availability of grid connection, cost of grid connection, battery storage
Determine the system capacity	Energy consumption patterns, shading analysis, grid connection options
Design the system	Solar panels, inverters, batteries, mounting structures, electrical wiring, protection systems, control systems
Conduct financial analysis	Cost of components, installation costs, maintenance costs, financial incentives and subsidies, payback period, return on investment
Prepare project documentation	Detailed design, project documentation, schematics, technical specifications, project budget
Obtain permits and approvals	Building permits, electrical permits, fire permits, local authority approvals
Installation and commissioning	Qualified installer, testing, commissioning
Maintenance and monitoring	Maintenance plan, monitoring plan, regular monitoring, regular maintenance

studies and practical practices examined in this chapter, offering a valuable resource for researchers, practitioners, and implementers in the field of PV systems.

Table 2 serves as an essential checklist, guiding the development of a PV system through each step of the project with clearly defined building blocks. It starts with site assessment, collecting crucial meteorological and topographical data, and progresses to load analysis to tailor the system design to the community's energy needs. The table includes steps for ensuring compliance with regulations, evaluating grid connectivity and system capacity, and encompasses detailed design considerations for the system's components. Financial analysis and documentation are emphasized for project sustainability and transparency. Furthermore, it highlights the need for obtaining permits and emphasizes the importance of regular maintenance and monitoring for the system's longevity. Overall, Table 2 is a structured tool that underpins the project's efficiency, compliance, and sustainability.

Detailed guidelines for the planning, design, and implementation of a PV system are given in Table 3, which encapsulates every critical step in the process, from initial assessment to long-term maintenance. It methodically breaks down each phase into specific building blocks and further into key segments or factors, ensuring a thorough and systematic approach. Starting with a site assessment that includes meteorological data and shading analysis, the table guides through energy consumption determination, adherence to local codes and regulations, grid connection assessment, and precise system capacity determination. It then delves into the

Table 3 A detailed guide for PV system planning, design, and implementation

Step	Building blocks	Segments/factors	Elements
Conduct a site assessment	Meteorological data	Average daily solar radiation levels	Monthly average solar radiation levels, hourly average solar radiation levels
		Average temperatures	Monthly average temperatures, hourly average temperatures
		Wind speeds	Monthly average wind speeds, hourly average wind speeds
	Location, topography	Latitude, longitude	Latitude, longitude (degrees)
		Altitude	Altitude (meters)
		Slope	Slope (degrees)
		Aspect	Aspect (degrees)
	Shading analysis	Shading from trees, buildings, etc.	Shading duration, shading intensity
Determine energy consumption	Load analysis	Type of appliances	List of appliances, appliance specifications
		Number of appliances	Number of appliances for each type
		Operating hours	Operating hours for each appliance
	Energy consumption patterns	Energy consumption of households and public facilities	Energy consumption per household, energy consumption per facility
Review local building codes and regulations	Building codes	Building design and construction standards	Building design standards, building construction standards
	Electrical codes	Electrical safety standards	Electrical safety standards, electrical safety codes
		Electrical wiring standards	Electrical wiring standards, electrical wiring codes
	Fire codes	Fire safety standards	Fire safety standards, fire safety codes
		Fire protection systems	Fire protection systems, fire protection codes
Assess grid connection options	Availability of grid connection	Access to the grid	Access to the grid, grid connection distance
	Cost of grid connection	Grid connection fees	Grid connection fees, interconnection fees
		Net metering fees	Net metering fees, net metering regulations
	Battery storage	Availability of battery storage	Battery storage availability, battery storage options
		Battery sizing	Battery sizing, battery capacity
Determine the system capacity	Energy consumption patterns	Energy demand of households and public facilities	Energy demand per household, energy demand per facility

(Continued)

Table 3 (continued)

Step	Building blocks	Segments/factors	Elements
	Shading analysis	Shaded hours	Shaded hours, shaded hours per day
		Shading impact on solar generation	Shading impact on solar generation, shading impact on energy generation
	Grid connection options	Grid connection availability	Grid connection availability, grid connection options
		Connection cost	Connection cost, interconnection cost
Design the system	Solar panels	Panel type	Panel type, panel specifications
		Panel size	Panel size, panel capacity
		Panel efficiency	Panel efficiency, panel performance data
	Inverters	Inverter type	Inverter type, inverter specifications
		Inverter size	Inverter size, inverter capacity
		Inverter efficiency	Inverter efficiency, inverter performance data
	Batteries	Battery type	Battery type, battery specifications
		Battery size	Battery size, battery capacity
Determine system components	System components	Racking system	Racking system type, racking system specifications
		Wiring and electrical components	Wiring type, electrical components specifications
		Monitoring system	Monitoring system type, monitoring system specifications
Establish a budget	Equipment cost	Solar panels cost	Solar panels cost per watt, solar panels cost per unit
		Inverters cost	Inverters cost per watt, inverters cost per unit
		Batteries cost	Batteries cost per watt, batteries cost per unit
		Racking system cost	Racking system cost per unit, racking system cost per installation
		Wiring and electrical components cost	Wiring and electrical components cost per unit, wiring and electrical components cost per installation
		Monitoring system cost	Monitoring system cost per unit, monitoring system cost per installation

(Continued)

Table 3 (continued)

Step	Building blocks	Segments/factors	Elements
Obtain necessary permits and approvals	Building permits	Building design approval	Building design approval, building construction approval
	Electrical permits	Electrical safety approval	Electrical safety approval, electrical wiring approval
	Fire permits	Fire safety approval	Fire safety approval, fire protection approval
Install the system	Installation process	Site preparation	Site preparation, site preparation time
		Equipment installation	Equipment installation, equipment installation time
		System commissioning	System commissioning, system commissioning time
Monitor and maintain the system	Monitoring	Energy generation monitoring	Energy generation monitoring, energy generation data
		System performance monitoring	System performance monitoring, system performance data
	Maintenance	Regular cleaning	Regular cleaning, cleaning frequency
		Battery replacement	Battery replacement, battery replacement frequency
		Panel replacement	Panel replacement, panel replacement frequency
		Inverter replacement	Inverter replacement, inverter replacement frequency

intricate design aspects of the system, such as solar panels, inverters, and batteries, alongside financial analysis to ensure cost-effectiveness and sustainability. Additionally, the table outlines the necessary steps for project documentation, permits and approvals, installation and commissioning, and crucial ongoing maintenance and monitoring activities. This guide is designed to be an invaluable resource for professionals and enthusiasts aiming to efficiently design and implement a PV system, ensuring all critical aspects are addressed for optimal functionality and compliance.

For a clearer understanding, a bill of quantities (BOQ) is proposed in Table 4, which provides a comprehensive list of all materials, equipment, and services required for the 10 kW PV system. This list includes the respective quantities, units, and costs of each item. The total projected cost of the system amounts to $5790.00. It is important to note that these figures are illustrative and actual costs may vary based on several factors, including location, equipment specifications, labor costs, and others.

Table 4 presents a detailed cost breakdown for the installation of a 10 kW PV system, itemizing each component required for a successful setup. The table lists items ranging from high-efficiency solar panels to essential building permits, specifying the quantity, unit price, and total price for each based on available prices on

Table 4 The proposed cost breakdown for 10 kW PV system installation and deployment

Item	Description	Quantity	Unit price	Total price	Qualities
Solar panels	High-efficiency solar panels	40	$0.70/W	$2800.00	Efficiency: 22%
Inverters	High-efficiency inverters	4	$0.60/W	$240.00	Maximum efficiency: 96%
Batteries	High-capacity batteries	4	$0.50/W	$200.00	Capacity: 100 Ah
Racking system	Aluminum racking system	1	$500.00/set	$500.00	Material: Aluminum
Wiring and electrical components	Copper wiring and electrical components	1	$300.00/set	$300.00	Wire gauge: 12 AWG
Monitoring system	Solar monitoring system	1	$200.00/set	$200.00	Monitoring accuracy: ±1%
Building permits	Building design and construction approval	1	$300.00/set	$300.00	
Electrical permits	Electrical safety and wiring approval	1	$200.00/set	$200.00	
Fire permits	Fire safety and protection approval	1	$150.00/set	$150.00	
Installation	Site preparation and equipment installation	1	$1000.00/set	$1000.00	
Monitoring and maintenance	Regular cleaning and monitoring	1	$300.00/set	$300.00	

the Internet. It also highlights the key qualities of these components, such as the efficiency of the solar panels and inverters and the capacity of the batteries. This thorough compilation not only provides a clear financial overview of the project but also ensures that all necessary elements for the installation, including safety and regulatory approvals, are accounted for. It is designed to offer a clear financial perspective to potential investors or implementers of a PV system, emphasizing the importance of each component in achieving an efficient and compliant solar energy solution. Note that these figures are for illustrative purpose only and actual costs may vary depending on specific circumstances such as site conditions and market variations.

Table 5 offers an exhaustive guide outlining the various elements and specifications involved in the installation of a solar panel system. It covers every component from the racking system and wiring to the monitoring system, detailing their types, specifications, and associated costs. Additionally, it includes the costs and types of solar panels, inverters, and batteries, which are critical to the system's operation. This table also addresses necessary approvals from local authorities for building design, construction, electrical safety, and fire protection, ensuring legal compliance. Furthermore, it delves into the practical aspects of site preparation, equipment installation, system commissioning, and ongoing maintenance activities like cleaning and component replacements. Each aspect is meticulously detailed, providing a clear and comprehensive view of what is required for a successful solar panel system installation, from initial planning stages to long-term maintenance and

Table 5 Components and specifications for PV system installation and maintenance

Elements	Explanation
Racking system type	Type of mounting structure to securely attach the solar panels to the ground or roof.
Racking system specifications	Dimensions, material, weight capacity, corrosion resistance, etc., of the racking system.
Wiring type	Type of cable and connectors used to transfer electricity from the panels to the inverters and battery bank.
Electrical components specifications	Voltage, current, and power rating of components such as fuses, switches, and circuit breakers.
Monitoring system type	Type of device used to monitor the performance of solar panels and the overall system.
Monitoring system specifications	Capacity, accuracy, and communication options of the monitoring system.
Solar panels cost per watt	Price per unit of power output of the solar panels.
Solar panels cost per unit	Total cost of a single solar panel.
Inverters cost per watt	Price per unit of power output of the inverters.
Inverters cost per unit	Total cost of a single inverter.
Batteries cost per watt	Price per unit of energy stored in the batteries.
Batteries cost per unit	Total cost of a single battery.
Racking system cost per unit	Cost of one racking system component.
Racking system cost per installation	Total cost of installing the racking system for the entire project.
Wiring and electrical components cost per unit	Cost of one electrical component such as a fuse or switch.
Wiring and electrical components cost per installation	Total cost of wiring and installing electrical components for the entire project.
Monitoring system cost per unit	Cost of one monitoring system component.
Monitoring system cost per installation	Total cost of installing the monitoring system for the entire project.
Building design approval	Approval from local authorities for the design of the building where the solar panels will be installed.
Building construction approval	Approval from local authorities for the construction of the building.
Electrical safety approval	Approval from local authorities for the electrical safety of the solar panel installation.
Electrical wiring approval	Approval from local authorities for the electrical wiring of the solar panel installation.
Fire safety approval	Approval from local authorities for the fire safety of the solar panel installation.
Fire protection approval	Approval from local authorities for the fire protection measures of the solar panel installation.
Site preparation	Preparation of the site where the solar panels will be installed, including leveling, excavation, and grading.
Site preparation time	Time required to prepare the site for solar panel installation.
Equipment installation	Installation of the solar panels, inverters, batteries, and monitoring system.

(continued)

Table 5 (continued)

Elements	Explanation
Equipment installation time	Time required to install all the equipment.
System commissioning	Final testing and adjustment of the solar panel system to ensure it is operating correctly.
System commissioning time	Time required to complete the commissioning process.
Energy generation monitoring	Monitoring of the amount of energy being generated by the solar panels.
Energy generation data	Data collected from the monitoring system on energy generation.
System performance monitoring	Monitoring of the overall performance of the solar panel system.
System performance data	Data collected from the monitoring system on system performance.
Regular cleaning	Regular cleaning of the solar panels to ensure maximum performance.
Cleaning frequency	The frequency of cleaning the solar panels to maintain optimal performance.
Battery replacement	The process of replacing the batteries in the system.
Battery replacement frequency	The frequency at which the batteries in the system need to be replaced.
Panel replacement	The process of replacing damaged or inefficient solar panels in the system.
Panel replacement frequency	The frequency at which solar panels in the system need to be replaced.
Inverter replacement	The process of replacing the inverter in the system.
Inverter replacement frequency	The frequency at which the inverter in the system needs to be replaced.

monitoring. This table is an essential tool for anyone involved in the planning, implementation, and upkeep of a solar panel system, offering valuable insights into the technical and administrative requirements of such projects.

4 Calculations and Estimation of the Stability of the PV System

Table 6 contains a detailed and systematic overview of the essential components and calculations required for the design and implementation of a PV system. It serves as a comprehensive guide, covering a wide range of factors from solar irradiation assessment to economic considerations, ensuring that each aspect of the PV system is thoroughly planned and optimized. The table not only lists the critical components necessary for a successful PV system deployment but also includes the respective calculation methods for each, providing valuable insights into the technical and financial planning processes. It can be counted an indispensable tool for professionals in the field of solar energy, offering a structured approach to designing efficient, safe, and economically viable PV systems.

Table 6 A managed methodology for a PV system planning and design of components [9, 27, 30, 44–51]

No.	Component	Calculation method
1	Solar irradiation	Estimate average daily solar radiation for the location.
2	PV size	Determine solar panel size based on daily electricity consumption and sunlight availability.
3	Battery sizing	Calculate battery bank size for energy storage during low sunlight or high consumption.
4	Inverter sizing	Determine inverter size for DC to AC power conversion.
5	System efficiency	Calculate total system efficiency, including losses from panels, batteries, and inverter.
6	Load analysis	Analyze the community's electricity consumption and determine peak load.
7	System voltage	Determine system voltage based on power needs, panel count, and installation space.
8	Panel orientation and tilt	Calculate optimal panel orientation and tilt for maximum energy generation.
9	Power generation	Estimate power generation based on panel size, orientation, tilt, and irradiation.
10	Economics	Conduct cost–benefit analysis including system component and maintenance costs.
11	Safety considerations	Review and comply with safety standards and regulations.
12	Short-circuit analysis	Calculate and analyze short-circuit.
13	Protection systems	Size fuses and circuit breakers based on short-circuit analysis.
14	Grounding	Design grounding system to meet electrical safety standards.
15	Load flow analysis	Use software to ensure balanced and stable power flow.
16	Maximum power point tracking (MPPT)	Design the MPPT system to optimize panel energy generation.
17	Maintenance considerations	Plan regular maintenance and component replacement.
18	Environmental considerations	Design the system to withstand local environmental conditions.
19	Racking system type	Select based on local regulations, conditions, and project requirements.
20	Racking system specifications	Specify size, type, and material based on project requirements.
21	Wiring type	Select based on project requirements, voltage drop, and ampacity.
22	Electrical components specifications	Specify components based on project needs.
23	Monitoring system type	Select based on data requirements and communication method.
24	Monitoring system specifications	Specify system based on project needs.
25	Solar panel cost per watt	Determine based on efficiency and capacity in watts.
26	Solar panel cost per unit	Determine based on size and quantity.
27	Inverter cost per watt	Determine based on efficiency and capacity in watts.
28	Inverter cost per unit	Determine based on size and quantity.
29	Battery cost per watt	Determine based on capacity in watt-hours.
30	Battery cost per unit	Determine based on size and quantity.

Table 7 plays as an essential tool in the design and optimization of photovoltaic (PV) systems. It meticulously categorizes key environmental and technical factors that influence PV system performance and viability. Each factor is identified with its respective measurement unit, and where applicable, additional supportive information is provided to enhance understanding. This table serves as a comprehensive reference, guiding professionals in accurately assessing and incorporating various parameters such as solar insolation, wind conditions, and geographical attributes. By considering these factors, one can ensure the effective design and efficient operation of PV systems, tailored to specific environmental and site conditions. This approach not only optimizes energy output but also contributes to the overall sustainability and economic feasibility of the PV project.

Table 7 Environmental and technical factors for PV system design [35, 45, 48, 49]

Name of the factor	Measuring unit of the factor	Supportive information if available
Distance-to-solar-noon	Radians	
Temperature	Degree Celsius	Daily average temperature
Wind direction	Degrees (0–360)	Daily average wind direction
Wind speed	Meters per second	Daily average wind speed
Sky cover	Five-step scale (0–4)	0: Totally clear; 4: Completely covered
Visibility	Kilometers	
Humidity	Percentage	
Average wind speed (period)	Meters per second	Average wind speed during the 3-hour period measured
Average pressure (period)	Mercury inches	Average barometric pressure during the 3-h period measured
Power generated		
Solar insolation	Kilowatt-hours per square meter	Average daily amount of solar radiation received
Latitude and longitude		Location of the proposed site and its impact on sunlight angle
Tilt and orientation of PV panels		Angle and direction relative to the sun, affecting energy production
Shading		Impact of obstructions like buildings and trees on sunlight access
Altitude		Elevation of the site, affecting air pressure and temperature
Ground reflectance		Reflectance of the ground surface affecting light reaching panels
Electrical load		Daily and seasonal electrical demand impacting system size and design
Grid availability		Accessibility of the electrical grid for connection
System maintenance		Frequency and cost of maintenance, including component replacement
Government regulations and incentives		Policies and incentives related to renewable energy in the area

In a PV system, the process of power flow is a crucial aspect, encompassing three primary stages [52]. Initially, power generation occurs when solar panels harness sunlight and convert it into DC electricity. This DC electricity is then processed in the second stage (power conditioning) where an inverter plays a pivotal role in transforming DC into AC. This AC electricity is suitable for household applications or can be fed into the electrical grid [53]. In the final stage (power delivery), the produced AC electricity is directly utilized by local loads, stored in batteries for later use, or distributed to the electrical grid. Notably, in grid-tied PV systems, any surplus energy generated can be transferred back to the grid, potentially earning credits for the system owner [54].

Optimization of PV projects involves a series of strategic considerations to enhance their efficiency and effectiveness. Site optimization is fundamental, involving the selection of the most suitable location while taking into account factors like solar insolation, shading, and local climate conditions. Panel optimization is another critical aspect, focusing on the choice of efficient panel technology and the strategic adjustment of their tilt and orientation to maximize energy capture [55]. Battery optimization also plays a significant role, emphasizing the importance of proper battery sizing, quality selection, and the incorporation of effective management systems. Load management is essential to align the PV system design with the community's daily and seasonal energy needs [56]. Regular maintenance and the use of high-quality components are vital in maintenance optimization, aiming to minimize long-term costs. Financial optimization involves exploring favorable financing options and tapping into available government incentives. Finally, operational optimization is achieved through the deployment of advanced monitoring and control systems, ensuring the efficient and smooth performance of the PV system [57].

5 Discussion and Lessons Learned

The implementation of PV systems, as observed in the case study of community electrification, presents several challenges. First, the variability of solar irradiation can affect the reliability of the power supply. To counteract this, the project incorporated robust battery storage solutions to ensure continuous power availability, especially during peak evening hours [12]. Another significant challenge was managing the cost of setting up and maintaining the system. This was partly mitigated by leveraging available financial incentives and subsidies, reducing the initial investment burden [11]. The technical complexity of designing a PV system to cater to fluctuating energy demands also posed a challenge [58]. Through detailed load analysis and adopting advanced solar forecasting models, such as the combined ARIMA and ANN models, the system's capacity was optimized to match the community's energy consumption patterns [13]. Finally, compliance with local building codes and regulations required careful navigation [59]. Ensuring legal compliance involved a thorough understanding of local standards and incorporating necessary adjustments in the system design, particularly in aspects like panel placement and electrical safety standards.

The PV system significantly contributed to reducing carbon emissions, aligning with global efforts to combat climate change. By replacing conventional fossil fuels with solar energy, the project also reduced air pollution, promoting better public health in the rural community [60]. Economically, the project demonstrated viability through a favorable payback period, enhanced by government subsidies. The reduction in energy costs for the community also translated into increased disposable income, potentially driving economic growth at the local level [61]. Socially, the project played a pivotal role in improving the quality of life by providing reliable electricity, which is essential for education, healthcare, and overall community development [62]. Electrification also opened doors to new technological advancements and digital connectivity for the community.

Future research will deal with integrating smart grid technologies to enhance the adaptability and efficiency of PV systems. Exploring emerging solar panel technologies with higher efficiencies and lower costs would also be valuable. Additionally, there is scope for developing more sophisticated energy storage solutions, which could further stabilize the power supply in varying weather conditions. Investigating the role of community participation in system maintenance and decision-making could also provide insights into enhancing project sustainability and local engagement. The lessons learned from the literature and practical implications can be listed as follows:

- Technical adaptability: The importance of designing PV systems that are adaptable to specific environmental conditions and user patterns cannot be overstated. Customized solutions yield better efficiency and user satisfaction.
- Economic viability: Balancing the upfront costs with long-term benefits is crucial. Accessing subsidies and incentives can significantly improve the economic feasibility of PV projects.
- Community engagement: Active involvement of the community in the planning and implementation phases enhances the system's acceptance and sustainability. Understanding and catering to local energy needs is vital.
- Regulatory compliance: Ensuring compliance with local regulations is essential for the smooth operation and legal validity of the project. Early engagement with regulatory bodies can streamline the process.
- Maintenance and monitoring: Regular maintenance and monitoring are key to sustaining the system's efficiency. This includes not just technical upkeep but also monitoring financial and environmental performance.
- Incorporating advanced technologies: The use of advanced forecasting and optimization models significantly improves the efficiency and reliability of PV systems.
- Environmental consciousness: PV systems should be designed with a strong emphasis on minimizing environmental impact, not just in terms of carbon emissions but also considering factors like land use and ecosystem preservation.

6 Conclusion

This chapter presents the significant role of solar PV systems in transforming energy markets. Solar energy, being clean and renewable, contributes effectively to energy security and climate change mitigation. The declining costs and scalability of solar technology enhance its accessibility and viability, making it a sustainable alternative to fossil fuels. Solar energy diversifies the energy mix, bolsters grid resilience, and facilitates the transition to a low-carbon economy, while addressing challenges like high energy costs and environmental degradation. The integration of advanced technologies, as highlighted in studies from 2013 to 2021, demonstrates the evolution in solar power forecasting. These innovations enhance the efficiency and reliability of solar power systems, contributing to more accurate solar radiation predictions and energy production efficiency.

The planning and design of PV systems involve a comprehensive analysis of various factors including site location, climatic conditions, electrical load, and energy consumption patterns. This chapter emphasizes the importance of considering local building codes, grid connection costs, financial incentives, and the technical specifications of solar panels and inverters. A balanced approach to these factors is crucial for achieving efficient, cost-effective, and environmentally sustainable solar installations. The case study of a rural community electrification project illustrates the practical application of these principles. It shows how a well-designed PV system can meet the community's energy needs, ensure financial viability, and contribute to sustainable development. This includes a systematic site assessment, load analysis, compliance review, and a comprehensive system design that accounts for future growth and peak load management. The financial analysis, including a detailed cost breakdown, plays a pivotal role in the viability of PV systems. This chapter also presents an illustrative Bill of Quantities for a 10 kW PV system, demonstrating the importance of each component in the overall cost and efficiency of the system. This chapter also covers the environmental and technical factors crucial for a PV system design. These include solar insolation, wind conditions, geographical attributes, and local climate conditions. By considering these factors, optimal energy output and system efficiency can be achieved.

Understanding the power flow process in PV systems is essential for optimization. The chapter explores the stages of power generation, conditioning, and delivery, emphasizing the importance of system optimization through site, panel, battery, and load management strategies. The lessons learned from this chapter highlight the need for continuous innovation in technology and forecasting methods, adherence to environmental and regulatory standards, and the adoption of a holistic approach to system design and implementation. Future directions include further advancements in technology, integration with smart grid systems, and exploring new financing and business models to make solar energy more accessible and affordable.

References

1. Sward, J.A., Siff, J., Gu, J., Zhang, K.M.: Strategic planning for utility-scale solar photovoltaic development – historical peak events revisited. Appl. Energy. **250**, 1292–1301 (2019). https://doi.org/10.1016/j.apenergy.2019.04.178
2. Lazaroiu, A.C., Gmal Osman, M., Strejoiu, C.-V., Lazaroiu, G.: A comprehensive overview of photovoltaic technologies and their efficiency for climate neutrality. Sustain. For. **15**, 16297 (2023). https://doi.org/10.3390/su152316297
3. Danish, M.S.S., Senjyu, T., Ibrahimi, A.M., Ahmadi, M., Howlader, A.M.: A managed framework for energy-efficient building. J. Build. Eng. **21**, 120–128 (2019). https://doi.org/10.1016/j.jobe.2018.10.013
4. Danish, M.S.S., Senjyu, T., Zaheb, H., Sabory, N.R., Ibrahimi, A.M., Matayoshi, H.: A novel transdisciplinary paradigm for municipal solid waste to energy. J. Clean. Prod. **233**, 880–892 (2019). https://doi.org/10.1016/j.jclepro.2019.05.402
5. Danish, M.S.S.: Exploring metal oxides for the hydrogen evolution reaction (HER) in the field of nanotechnology. RSC Sustain. **1**, 2180–2196 (2023). https://doi.org/10.1039/D3SU00179B
6. Ahmadi, M., Lotfy, M.E., Danish, M.S.S., Ryuto, S., Yona, A., Senjyu, T.: Optimal multi-configuration and allocation of SVR, capacitor, centralised wind farm, and energy storage system: a multi-objective approach in a real distribution network. IET Renew. Power Gener. **13**, 762–773 (2019). https://doi.org/10.1049/iet-rpg.2018.5057
7. Adewuyi, O.B., Shigenobu, R., Senjyu, T., Lotfy, M.E., Howlader, A.M.: Multiobjective mix generation planning considering utility-scale solar PV system and voltage stability: Nigerian case study. Electr. Power Syst. Res. **168**, 269–282 (2019). https://doi.org/10.1016/j.epsr.2018.12.010
8. Vangala, S., Hung, K., South, D.: Revisiting the Biden Administration's approach to climate change. Climate Energy. **39**, 1–12 (2022). https://doi.org/10.1002/gas.22299
9. Ahmadi, M., Jafari, S., Quraishi, H., Zaheb, H., Danish, M.S.S., Senjyu, T.: Partial deployment of rooftop solar and an energy efficiency approach for decreasing grid dependency. In: 2021 International Conference on Science Contemporary Technologies (ICSCT), pp. 1–8. IEEE, Dhaka (2021). https://doi.org/10.1109/ICSCT53883.2021.9642529
10. Ahmed, R., Sreeram, V., Mishra, Y., Arif, M.D.: A review and evaluation of the state-of-the-art in PV solar power forecasting: techniques and optimization. Renew. Sust. Energ. Rev. **124**, 109792 (2020). https://doi.org/10.1016/j.rser.2020.109792
11. Shams, S., Danish, M.S.S., Sabory, N.R.: Solar energy market and policy instrument analysis to support sustainable development. In: Danish, M.S.S., Senjyu, T., Sabory, N.R. (eds.) Sustainability Outreach in Developing Countries, pp. 113–132. Springer, Singapore (2021). https://doi.org/10.1007/978-981-15-7179-4_8
12. Danish, S.M.S., Ahmadi, M., Danish, M.S.S., Mandal, P., Yona, A., Senjyu, T.: A coherent strategy for peak load shaving using energy storage systems. J. Energy Storage. **32**, 101823 (2020). https://doi.org/10.1016/j.est.2020.101823
13. Danish, M.S.S., Nazari, Z., Senjyu, T.: AI-coherent data-driven forecasting model for a combined cycle power plant. Energy Convers. Manag. **286**, 117063 (2023). https://doi.org/10.1016/j.enconman.2023.117063
14. Naranjo Palacio, S., Valentine, K.F., Wong, M., Zhang, K.M.: Reducing power system costs with thermal energy storage. Appl. Energy. **129**, 228–237 (2014). https://doi.org/10.1016/j.apenergy.2014.04.089
15. Driesen, J., Katiraei, F.: Design for distributed energy resources. IEEE Power Energy Mag. **6**, 30–40 (2008). https://doi.org/10.1109/MPE.2008.918703
16. Alarcon-Rodriguez, A., Ault, G., Galloway, S.: Multi-objective planning of distributed energy resources: a review of the state-of-the-art. Renew. Sust. Energ. Rev. **14**, 1353–1366 (2010). https://doi.org/10.1016/j.rser.2010.01.006

17. Sarzynski, A., Larrieu, J., Shrimali, G.: The impact of state financial incentives on market deployment of solar technology. Energy Policy. **46**, 550–557 (2012). https://doi.org/10.1016/j.enpol.2012.04.032
18. Nikolakakis, T., Fthenakis, V.: The optimum mix of electricity from wind- and solar-sources in conventional power systems: evaluating the case for New York state. Energy Policy. **39**, 6972–6980 (2011). https://doi.org/10.1016/j.enpol.2011.05.052
19. Denholm, P., Hand, M.: Grid flexibility and storage required to achieve very high penetration of variable renewable electricity. Energy Policy. **39**, 1817–1830 (2011). https://doi.org/10.1016/j.enpol.2011.01.019
20. Adli, H.K., Husin, K.A.K., Hanafiah, N.H.M., Remli, M.A., Ernawan, F., Wirawan, P.W.: Forecasting and analysis of solar power output from integrated solar energy and IoT system. In: 2021 5th International Conference on Informatics and Computational Sciences (ICICoS), pp. 222–226. IEEE, Semarang (2021). https://doi.org/10.1109/ICICoS53627.2021.9651831
21. Qing, X., Niu, Y.: Hourly day-ahead solar irradiance prediction using weather forecasts by LSTM. Energy. **148**, 461–468 (2018). https://doi.org/10.1016/j.energy.2018.01.177
22. Sivaneasan, B., Yu, C.Y., Goh, K.P.: Solar forecasting using ANN with fuzzy logic pre-processing. Energy Procedia. **143**, 727–732 (2017). https://doi.org/10.1016/j.egypro.2017.12.753
23. Gutierrez-Corea, F.-V., Manso-Callejo, M.-A., Moreno-Regidor, M.-P., Manrique-Sancho, M.-T.: Forecasting short-term solar irradiance based on artificial neural networks and data from neighboring meteorological stations. Sol. Energy. **134**, 119–131 (2016). https://doi.org/10.1016/j.solener.2016.04.020
24. Abuella, M., Chowdhury, B.: Solar power forecasting using artificial neural networks. In: 2015 North American Power Symposium (NAPS), pp. 1–5. IEEE, Charlotte (2015). https://doi.org/10.1109/NAPS.2015.7335176
25. Dahmani, K., Dizene, R., Notton, G., Paoli, C., Voyant, C., Nivet, M.L.: Estimation of 5-min time-step data of tilted solar global irradiation using ANN (artificial neural network) model. Energy. **70**, 374–381 (2014). https://doi.org/10.1016/j.energy.2014.04.011
26. Marquez, R., Pedro, H.T.C., Coimbra, C.F.M.: Hybrid solar forecasting method uses satellite imaging and ground telemetry as inputs to ANNs. Sol. Energy. **92**, 176–188 (2013). https://doi.org/10.1016/j.solener.2013.02.023
27. Danish, M.S.S.: A framework for modeling and optimization of data-driven energy systems using machine learning. IEEE Trans. Artif. Intell.., 1–10 (2023). https://doi.org/10.1109/TAI.2023.3322395
28. Danish, M.S.S., Elsayed, M.E.L., Ahmadi, M., Senjyu, T., Karimy, H., Zaheb, H.: A strategic-integrated approach for sustainable energy deployment. Energy Rep. **6**, 40–44 (2020). https://doi.org/10.1016/j.egyr.2019.11.039
29. Danish, M.S.S., Senjyu, T.: AI-enabled energy policy for a sustainable future. Sustain. For. **15**, 7643 (2023). https://doi.org/10.3390/su15097643
30. Danish, M.S.S., Yona, A., Senjyu, T.: Pre-design and life cycle cost analysis of a hybrid power system for rural and remote communities in Afghanistan. J. Eng. IET. **2014**, 438–444 (2014). https://doi.org/10.1049/joe.2014.0172
31. Karatas, Y., Yilmaz, D.: Experimental investigation of the microclimate effects on floating solar power plant energy efficiency. Clean Techn. Environ. Policy. **23**, 2157–2170 (2021). https://doi.org/10.1007/s10098-021-02122-y
32. Zhong, Q., Nelson, J.R., Tong, D., Grubesic, T.H.: A spatial optimization approach to increase the accuracy of rooftop solar energy assessments. Appl. Energy. **316**, 119128 (2022). https://doi.org/10.1016/j.apenergy.2022.119128
33. Lee, S., Kim, J., Jang, D.: Analysis of major temporary electrical equipment consumption and usage patterns in educational buildings: case study. Sustain. For. **14**, 10783 (2022). https://doi.org/10.3390/su141710783

34. Niskanen, J., Rohracher, H.: A politics of calculation: negotiating pathways to zero-energy buildings in Sweden. Technol. Forecast. Soc. Chang. **179**, 121630 (2022). https://doi.org/10.1016/j.techfore.2022.121630
35. Das, B.K., Alotaibi, M.A., Das, P., Islam, M.S., Das, S.K., Hossain, M.A.: Feasibility and techno-economic analysis of stand-alone and grid-connected PV/wind/diesel/batt hybrid energy system: a case study. Energ. Strat. Rev. **37**, 100673 (2021). https://doi.org/10.1016/j.esr.2021.100673
36. Zander, K.K., Simpson, G., Mathew, S., Nepal, R., Garnett, S.T.: Preferences for and potential impacts of financial incentives to install residential rooftop solar photovoltaic systems in Australia. J. Clean. Prod. **230**, 328–338 (2019). https://doi.org/10.1016/j.jclepro.2019.05.133
37. Masili, M., Ventura, L.: Local tilt optimization of photovoltaic solar panels for maximum radiation absorption. Int. J. Photoenergy. **2019**, e3254780 (2019). https://doi.org/10.1155/2019/3254780
38. Stephens, G., Dieterle, C., Hossain, E., Bayindir, R.: Feasibility study of grid-connected solar plant: an in-depth analysis of system modeling and proper technology selection. Int. J. Electr. Eng. Educ. **60**, 159–175 (2023). https://doi.org/10.1177/0020720920928543
39. Madhu, M.C., Narayana, K.B., Kishore, J.K.: Hybrid DC-AC homes with roof top solar power. E3S Web Conf. **239**, 00017 (2021). https://doi.org/10.1051/e3sconf/202123900017
40. Aboagye, B., Gyamfi, S., Ofosu, E.A., Djordjevic, S.: Investigation into the impacts of design, installation, operation and maintenance issues on performance and degradation of installed solar photovoltaic (PV) systems. Energy Sustain. Dev. **66**, 165–176 (2022). https://doi.org/10.1016/j.esd.2021.12.003
41. Alhawsawi, E.Y., Habbi, H.M.D., Hawsawi, M., Zohdy, M.A.: Optimal design and operation of hybrid renewable energy systems for Oakland University. Energies. **16**, 5830 (2023). https://doi.org/10.3390/en16155830
42. Wang, W., Tu, J., Xu, H., Qi, F., Tavasoli, M., Su, Z.: Tow-sectional optimized thermodinamical cycle using different renewable energies including geothermal and biogas to produce stable productions. Renew. Energy. **220**, 119617 (2024). https://doi.org/10.1016/j.renene.2023.119617
43. Garraín, D., Herrera, I., Rodríguez-Serrano, I., Lechón, Y., Hepbasli, A., Araz, M., Biyik, E., Yao, R., Shahrestani, M., Essah, E., Shao, L., Rico, E., Lechón, J.L., Oliveira, A.C.: Sustainability indicators of a naturally ventilated photovoltaic façade system. J. Clean. Prod. **266**, 121946 (2020). https://doi.org/10.1016/j.jclepro.2020.121946
44. Hernández-Callejo, L., Gallardo-Saavedra, S., Alonso-Gómez, V.: A review of photovoltaic systems: design, operation and maintenance. Sol. Energy. **188**, 426–440 (2019). https://doi.org/10.1016/j.solener.2019.06.017
45. Al-Najideen, M.I., Alrwashdeh, S.S.: Design of a solar photovoltaic system to cover the electricity demand for the faculty of engineering- Mu'tah University in Jordan. Resour. Effici. Technol. **3**, 440–445 (2017). https://doi.org/10.1016/j.reffit.2017.04.005
46. Moussa, I., Khedher, A.: Chapter 4 - an experimental test bench for emulating the standard characteristics of photovoltaic (PV) systems. In: Jeguirim, M. (ed.) Renewable Energy Production and Distribution, pp. 107–135. Academic Press (2022). https://doi.org/10.1016/B978-0-323-91892-3.00011-X
47. Rawat, R., Kaushik, S.C., Lamba, R.: A review on modeling, design methodology and size optimization of photovoltaic based water pumping, standalone and grid connected system. Renew. Sust. Energ. Rev. **57**, 1506–1519 (2016). https://doi.org/10.1016/j.rser.2015.12.228
48. Oshiro, T., Nakamura, H., Imataki, M., Sakuta, K., Kurokawa, K.: Practical values of various parameters for PV system design. Sol. Energy Mater. Sol. Cells. **47**, 177–187 (1997). https://doi.org/10.1016/S0927-0248(97)00038-X
49. Iringová, A., Kovačic, M.: Design and optimization of photovoltaic systems in a parking garage – a case study. Trans. Res. Procedia. **55**, 1171–1179 (2021). https://doi.org/10.1016/j.trpro.2021.07.097

50. Ludin, G.A., Matayoshi, H., Danish, M.S.S., Yona, A., Senjyu, T.: Hybrid PV/wind/diesel based distributed generation for an off-grid rural village in Afghanistan. J. Energy Power Eng. **11**, 85–94 (2017). https://doi.org/10.17265/1934-8975/2017.02.003
51. Yaqobi, M.A., Matayoshi, H., Danish, M.S.S., Urasaki, N., Howlader, A.M., Senjyu, T.: Control and energy management strategy of standalone DC microgrid cluster using PV and battery storage for rural application. Int. J. Power Energy Res. **2**, 53–68 (2018). https://doi.org/10.22606/ijper.2018.24001
52. Indu Rani, B., Saravana Ilango, G., Nagamani, C.: Power flow management algorithm for photovoltaic systems feeding DC/AC loads. Renew. Energy. **43**, 267–275 (2012). https://doi.org/10.1016/j.renene.2011.11.035
53. Castro, L.M., Rodríguez-Rodríguez, J.R., Martin-del-Campo, C.: Modelling of PV systems as distributed energy resources for steady-state power flow studies. Int. J. Electr. Power Energy Syst. **115**, 105505 (2020). https://doi.org/10.1016/j.ijepes.2019.105505
54. Nazir, R., Kanada, K., Coveria, P.: Optimization active and reactive power flow for PV connected to grid system using Newton Raphson method. Energy Procedia. **68**, 77–86 (2015). https://doi.org/10.1016/j.egypro.2015.03.235
55. Singh, S.K., Lohani, B., Arora, L., Choudhary, D., Nagarajan, B.: A visual-inertial system to determine accurate solar insolation and optimal PV panel orientation at a point and over an area. Renew. Energy. **154**, 223–238 (2020). https://doi.org/10.1016/j.renene.2020.02.107
56. Mubaarak, S., Zhang, D., Wang, L., Mohan, M., Kumar, P.M., Li, C., Zhang, Y., Li, M.: Efficient photovoltaics-integrated hydrogen fuel cell-based hybrid system: energy management and optimal configuration. J. Renew. Sustain. Energy. **13**, 013502 (2021). https://doi.org/10.1063/1.5141932
57. Yuan, J., Yuan, W., Yuan, J., Liu, Z., Liao, J., Ou, X.: Policy recommendations for distributed solar PV aiming for a carbon-neutral future. Sustain. For. **15**, 3005 (2023). https://doi.org/10.3390/su15043005
58. Danish, M.S.S., Sabory, N.R., Danish, S.M.S., Ludin, G.A., Yona, A., Senjyu, T.: An open-door immature policy for rural electrification: a case study of Afghanistan. Int. J. Sustain. Green Energy. **6**, 8–13 (2016). https://doi.org/10.11648/j.ijrse.s.2017060301.12
59. Danish, S.M.S., Zaheb, H., Sabori, N.R., Karimy, H., Faiq, A.B., Fedayi, H., Senjyu, T.: The road ahead for municipal solid waste management in the 21st century: a novel-standardized simulated paradigm. In: IOP Conference Series: Earth and Environmental Science, p. 012009. IOP Publishing, Seoul (2019). https://doi.org/10.1088/1755-1315/291/1/012009
60. Martínez Torres, J., Pastor Pérez, J., Sancho Val, J., McNabola, A., Martínez Comesaña, M., Gallagher, J.: A functional data analysis approach for the detection of air pollution episodes and outliers: a case study in Dublin, Ireland. Mathematics. **8**, 225 (2020). https://doi.org/10.3390/math8020225
61. Adalı, Z., Danish, M.S.S.: Investigation of the nexus between the electricity consumption and the ecological footprint. In: Dinçer, H., Yüksel, S. (eds.) Circular Economy and the Energy Market: Achieving Sustainable Economic Development Through Energy Policy, pp. 79–89. Springer International Publishing, Cham (2022). https://doi.org/10.1007/978-3-031-13146-2_7
62. Danish, M.S.S., Senjyu, T., Urasaki, N., Rahmany, N.A., Ershad, A.M., Sabory, N.R., Zarabi, K., Anwarzai, M.A., Karimy, H., Zaheb, H.: Developing nations as a foremost allotment for achieving 2030 SDGs – a case study. J. Eng. Technol. Revolut. **2**, 1–10 (2021). https://doi.org/10.37357/1068/jetr.2.1.01

Index

M. S. S. Danish (ed.), *Unified Vision for a Sustainable Future*,
https://doi.org/10.1007/978-3-031-53574-1